AF478430

Stability theory and its applications
to structural mechanics

Monographs and textbooks on mechanics of solids and fluids

editor-in-chief: G. Æ. Oravas

Mechanics of elastic stability

editor: H. Leipholz

1. H. LEIPHOLZ
   Theory of elasticity

2. L. LIBRESCU
   The elasto-statics and kinetics of anisotropic and heterogeneous shell-type
   structures

3. C. L. DYM
   Stability theory and its applications to structural mechanics

4. K. HUSEYIN
   Nonlinear theory of elastic stability

Stability theory and its applications to structural mechanics

Clive L. Dym

*Bolt Beranek and
Newman Inc.
Waltham, Massachusetts*

NOORDHOFF INTERNATIONAL PUBLISHING – LEYDEN

ISBN: 90 286 009 49

Library of Congress Catalog Card Number: 73-94269

Printed in the Netherlands

Contents

Contents

Nothing is required and nothing will avail,
except a little, a very little, clear thinking.
John Maynard Keynes

Preface

This book represents in large part the material I have taught in graduate courses on structural stability theory and its applications at the State University of New York at Buffalo and at Carnegie-Mellon University. It represents my view that structural stability should be taught as a mathematical discipline which examines the stability of motion. Thus, if static approaches are to be used, they must be justified within the framework of a dynamic description. Further, if linearizations are to be used, they should be consistently developed, and they should not be highly dependent on free-body diagrams or physical intuition. Finally, the subject matter should be as open-ended as is possible in a short (one semester) introduction, so that applications to and commonalities with other disciplines are not made unnecessarily obscure.

The restriction to limit the length of the book to material that could be reasonably covered in one semester is self-imposed, as the aim in producing this work was to write a short, self-contained, textbook. Thus I have not chosen, for example, to include any work on shell stability. Nor are the examples given exhaustive in their coverage. However, I believe that the essence of a modern viewpoint of structural stability theory is contained herein, together with a sufficient number of examples and illustrations.

Needless to say, I have relied heavily on previously published work, including research papers, monographs and textbooks. Consequently I have tried to be meticulous in citing the works from which I have learned, and I hope that I've not erred on the side of omission.

I would like to thank the editor of this series, Professor H. H. E. Leipholz, for his understanding and his encouragement. And I would also like to pay tribute to Miss Margaret Hall, who so ably typed the manuscript.

Finally, I would like to dedicate this book to my two 'Monsters', Jordana and Miriam, and to the memory of their late grandfather, Irving Gelb. Irv always got a kick out of my writing—'You get *paid* to do this?'— and so I think it is fitting to dedicate this volume to his memory.

Pittsburgh, Pennsylvania   *Clive L. Dym*
May 1974

1

General notions; agenda

The study of stability, for someone who is interested in the study of mechanics and mathematics, is both a fascinating and a challenging task. To begin with, we all think that we have an intuitive notion about what stability means, which is generally based on a kinematic criterion used in the development of the dynamics of a rigid body. Thus a ball on the top of a hill, when given a small displacement from its resting position, will continue to roll of its own accord, and thus is held to have been in an *unstable* position. Conversely, if a ball in a trough is lightly pushed from its resting position, it will tend to return to the bottom of the trough, and thus will be accounted in a *stable* position. However, this 'definition' is clearly inadequate to deal with a ball on a flat plane. If such a ball is lightly pushed it will move a little, perhaps, but it will neither return to its starting position, nor will it continue to roll indefinitely, of its own accord.

Another way that the aforementioned problem may be looked at is to observe that in the unstable case the center of gravity of the ball moved downward with respect to a fixed point at the top of the hill, in the stable case the disturbance attempted to raise the center of gravity of the ball with respect to the trough, while for the ball on the plane the relative (vertical) position of the center of gravity remained unchanged. Thus the stability of the ball on the plane should be termed *neutral* because the position of its center of gravity is neither lowered or raised.

This second viewpoint suggests that the concept of *motion* should be introduced as part of a discussion of stability, for if we are to examine changes (or possible changes) of position, we are implying some (non-static) movement. In fact, to those of us familiar with the usual eigenvalue formulation of the Euler column problem, it must seem somewhat artificial to discuss adjacent states of equilibrium without discussing the requisite dynamic process of moving from one state to another. The concept of stability being a dynamic phenomenon is one that we shall wish to emphasize, even as we lay the foundation for the (static) minimum energy criterion for conservative systems.

There are many other related phenomena that we will attempt to discuss. For example, the simple harmonic oscillator

$$\ddot{y} + y = 0 \tag{1.1}$$

or

$$\dot{x}_1 = x_2$$
$$\tag{1.2}$$
$$\dot{x}_2 = -x_1$$

has a well-known solution,

$$y = y(0) \cos t + \dot{y}(0) \sin t \tag{1.3}$$

Is this solution stable?

We cannot answer that question without defining what it means for the solution of a differential equation to be stable. Does it mean, for example, that the solution remains bounded for all time? Or, does stability require that $y(t) \to 0$ as $t \to \infty$. Obviously this requires that we define stability, or different types of stability, in some reasonably precise fashion.

Another area of some interest to us is the influence of nonlinearities on stability predictions. To extend the above example, consider the differential equation system

$$\dot{x}_1 = x_2 - \varepsilon(x_1^2 + x_2^2)$$

$$\dot{x}_2 = -x_1 - \varepsilon(x_1^2 + x_2^2) \tag{1.4}$$

which is rather nonlinear, even for the small values of the parameter ε in which we will be interested in the sequel. Clearly, if $\varepsilon = 0$, the system (1.4) reduces to the simple harmonic oscillator. Now a number of questions ought to be considered. If ε is truly small, so that the effect of the nonlinear terms is small, will the stability behavior of equations (1.4) be reasonably similar to that of the oscillator? If we further assume — not a very big assumption, by the way — that the nonlinearity makes the system (1.4) intractable with respect to our usual techniques for differential equations, then are there other ways that we can elicit information without actually solving the equations?

Another set of topics that we shall be required to take up are questions of *discrete* and *continuous systems*, i.e., questions of when a stability criterion developed for discrete systems may be applied to continuous systems, and when may a continuous system be replaced by some combination of discrete systems.

Finally, and perhaps of most importance, we shall be interested in the applications of the stability theory that we will develop to problems that are both interesting and relevant.

It will be our aim, in this book, to heuristically develop and apply the theory of stability of motion. Although we shall endeavor to be correct and reasonably precise, we intend to be informal rather than rigorous in our presentation.

We shall devote our early attention to ordinary differential equations, particularly the geometric theory. In this context we shall introduce various definitions of stability, the concept of variational or perturbed equations, measurement of 'distances', and the fundamental ideas of phase plane analysis. This will allow for the introduction of linear equilibrium points for a single-degree-of-freedom system, and the introduction of a theorem on nonlinearity. Then we shall introduce some first integrals and energy definitions, so as to present the theorems of Dirichlet and Lyapunov for the first time. We will also mention here Poincaré's limit cycles.

We shall then proceed to more general linear systems, introducing the stability criteria of Nyquist and of Routh and Hurwitz. Then after some further discussion and a theorem on the role of nonlinearities, we present the direct method of Lyapunov. With the aid of a Lyapunov function we can present the major stability results for a discrete system, including Lyapunov stability and instability theorems, the instability theorem of Chetayev, and certain further generalizations. Finally we recast some of these theorems in terms of the canonical equations of mechanics.

The next step is the extension of the Lyapunov approach to continuous systems governed by partial differential equations. In this context we shall introduce Lyapunov functionals, and state some principal results. Then, in turn, we shall be able to develop the minimum energy principle for a continuous (conservative) system.

We then proceed to applications chosen largely from the field of elastic structural systems. (Note that all the aforementioned developments will be threaded with examples.) These examples follow an exposition of the relevant nonlinear structural theories, and will include the buckling and postbuckling of columns, plates, and arches. Also covered will be the asymptotic approach to postbuckling behavior developed by Koiter. Finally some applications of Lyapunov functionals to dynamic stability problems will be displayed.

In sum the book has three principal parts: 1) The development of the Lyapunov direct method and of the minimum energy principle for discrete systems; 2) Analogous development of the Lyapunov functional method and of the minimum energy approach for continuous systems; 3) Applications.

2

Some geometric theory

2.1 Definitions of stability

We shall be interested in examining, in this chapter, features of the system of differential equations represented as

$$\dot{x}_i = X_i(x_1, x_2 \ldots x_{2n}, t), \; i = 1, 2 \ldots, 2n \tag{2.1}$$

where the x_i and X_i may also be regarded as components of the $2n$-dimensional vectors x and X. Note that the parameter for time, t, appears explicitly in equation (2.1), which implies that system is *non-autonomous*. Most of our applications will be to *autonomous* systems, i.e.,

$$\dot{x}_i = X_i(x_1, x_2 \ldots x_{2n}), i = 1, 2 \ldots 2n \tag{2.2}$$

or

$$\dot{x} = X(x) \tag{2.3}$$

We now define a Euclidian space of dimension $2n$, defined in terms of the variables x_i, as the *phase space* in which the motion described by equations (2.2) or (2.3) can be geometrically represented. For the non-autonomous system (2.1) we should have to define a $(2n + 1)$-dimensional *motion space* of (x, t). Also note that we have not stated any restrictions regarding the linearity (or nonlinearity) of equations (2.2) or (2.3).

From a mathematical viewpoint, the differential equations (2.2) will have unique solutions for a given set of initial conditions if the functions X_i are *Lipschitzian*. A Lipschitz condition in a domain D requires that

$$| X_i (x_1', x_2' \ldots x_{2n}') - X_i (x_1'', x_2'' \ldots x_{2n}'') |$$

$$\leq A \sum_{i=1}^{2n} | x_i' - x_i'' | \tag{2.4}$$

for all x', x'' in the domain D, for some value of the Lipschitz constant A. It is sufficient for our purposes to note that the Lipschitz condition is slightly stronger than a continuity condition, but not quite as restrictive as requiring differentiability. We shall assume that all our equations are Lipschitzian, so that by a fundamental theorem, the equations have unique solutions (see Tricomi [1] [1]).

The curves of x projected in phase space are called *trajectories*. Further, if $t = t_0$ is the time at which the trajectories are begun, the trajectories are termed *positive* or *negative* according to whether $t \geq t_0$ or $t \leq t_0$. A point where X vanishes is termed an *equilibrium* or *singular* point, while all other points are *regular*.

Now in order to define whether or not a solution is stable, or not, it is (intuitively, at least) clear that this definition will involve in part a measure of distance, so that we can define a solution x being close to, or far from, some curve or set of points. Thus we introduce a *Euclidian norm* or *metric*

$$|| x || = \left(\sum_{i=1}^{2n} x_i^2 \right)^{1/2} = (x^T x)^{1/2} \tag{2.5}$$

where x^T is the transpose of the column vector x, and is thus a row vector. Then we can say that a curve $x(t)$ in the phase space is *bounded* if $|| x(t) || < r$, where r is a given constant, for all time. Note that this inequality defines a spherical region of radius r in the phase space.

We may relate our stability problems from mechanics to the differential equation (2.2) in the following way. We may assume without loss of generality that the origin $x = 0$ is an (isolated) equilibrium point, where, also, $X(0, t) = 0$. We then wish to inquire, for the dynamical system

$$\dot{x} = X(x, t)$$

whether a slight perturbation from equilibrium will produce motion that

---

1 Numbers in brackets refer to References at the ends of the Chapters.

will return to equilibrium, or remain in a small neighborhood of equilibrium, or diverge altogether from equilibrium.

The definitions of stability according to Lyapunov, which we have just hinted at, are customarily given in terms of the *origin* of phase space, or the *null vector* $x = 0$, rather than in terms of the *disturbance* $x(t)$. Thus, denoting a particular (disturbance) integral curve at some $t_0 > 0$ by $x(t_0) = x_0$, we introduce the following definitions [2, 3, 4]:

1) The origin is *Lyapunov stable* if for any arbitrary positive ε and time t_0, there exists a $\delta = \delta(\varepsilon, t_0)$ such that if

$$\| x_0 \| < \delta \tag{2.6}$$

then

$$\| x(t) \| < \varepsilon, \ t_0 < t < \infty \tag{2.7}$$

is implied. If $\delta = \delta(\varepsilon)$ only, the stability is *uniform*.

2) The origin is *asymptotically stable* if it is both Lyapunov stable and such that

$$\lim_{t \to \infty} \| x(t) \| = 0 \tag{2.8}$$

3) The origin is *unstable* if for arbitrarily small δ and any time t_0 such that

$$\| x_0 \| < \delta \tag{2.9}$$

we have another time t_1

$$\| x(t_1) \| > \varepsilon, \ t_0 < t_1 < \infty \tag{2.10}$$

Thus we see that Lyapunov stability requires — by definition — that the solution always remains within a sphere of radius ε of the origin. Asymptotic stability further requires that the solution return to equilibrium as time becomes infinite. Instability occurs if the sphere of radius ε is reached. Note also that for autonomous systems, the role played by t_0 is essentially trivial as time doesn't appear in $X(x)$, which indicates that the Lyapunov stability is always uniform.

Now the definition of Lyapunov stability is rather restrictive, because it is easy to see, for example, that periodic motion may be excluded. If we consider a closed trajectory in phase space, then we can define orbital stability and asymptotic orbital stability with respect to that closed trajectory. If x' is a point[1] on the closed trajectory and x is any point near the trajectory, then we can define the distance from x to the closed trajectory C as

$$d(x, C) = \inf \{d(x, x'): x' \in C\} \tag{2.11}$$

where

$$d(x, x') = || x - x' || \tag{2.12}$$

and inf is the infimum – or smallest value – of the distance from x to x'. Then we can introduce the following definitions:

4)   The closed trajectory C is said to be *Poincaré stable* or *orbitally stable* if, given a positive ε, there exists a positive δ such that every solution $x(t)$ for which $x(t_0) = x_0$, and where

$$d(x_0, C) < \delta \tag{2.13}$$

has the property that

$$d(x, C) < \varepsilon, t \geq t_0 \tag{2.14}$$

This implies that the solution $x(t)$ stays near the closed trajectory $x'(t)$. Also, we have the definition

5)   The closed trajectory is said to be *asymptotically orbitally stable* if it is both orbitally stable and if there is a positive ε_0 such that for every solution $x(t)$ for which $x(t_0) = x_0$, the inequality

$$d(x_0, C) < \varepsilon_0 \tag{2.15}$$

implies

$$\lim_{t \to \infty} d(x(t), C) = 0 \tag{2.16}$$

---

1 Recognize that we are actually identifying a point by its vector.

Thus, in this definition, the solution $x(t)$ becomes coincident with the closed trajectory as time becomes infinite.

There are, obviously, other possible definitions, e.g., that of Lagrange, which requires only simple boundedness for $x(t)$ for $t > t_0$. However, this definition has not been found very useful.

Another stability definition that might find application in control problems, or other areas where only finite time intervals are of interest, is *stability over a finite time interval*. Here we require the conditions of Definition 1 above, excepting only that (see equation (2.7))

$$\| x(t) \| < \varepsilon, \, t_0 < t < kt_0$$

where k is some constant.

Another definition of stability is that of *practical stability*, advanced by LaSalle and Lefschetz [3, pp. 121-126]. Here the proposition is one of considering a fundamental system (2.1) and a perturbed system

$$\dot{x} = X(x, t) + p(x, t)$$

Assume that $x^*(t, x_0)$ is a solution of the perturbed system, with $x^*(t_0, x_0) = x_0$, with the perturbations satisfying $\| p(x, t) \| < \delta$ for all $t > 0$ and all x. Then for each such $p(x, t)$, if $x^*(t, x_0)$ starts within an acceptable set of initial states, and ends within an acceptable set of final states, the origin is said to be *practically stable*.

Before getting into some elementary phase plane analyses, it is worth reiterating one point that was glossed over. We had assumed earlier that the origin could be described by $x = 0$, and $X(0, t) = 0$. Let us suppose, however, that we had a particular solution $\phi(t)$, or $\phi_i(t)$, to equation (2.1). Then we could introduce the *perturbed* or *variational* motion $x(t)$ in the neighborhood of the unperturbed motion $\phi(t)$, defined by the *perturbations*

$$\dot{y}(t) = x(t) - \phi(t) \tag{2.17}$$

Then, from equations (2.1) and (2.17) we have,

$$\dot{y}(t) = X(y + \phi, t) - X(\phi, t) = Y(y, t) \tag{2.18}$$

Notice now that a trivial solution $y = 0$ exists as an equilibrium point for equation (2.18) since $Y(0, t) = 0$. Thus we have vindicated our earlier assertion that we lose no generality by assuming the existence of such a trivial equilibrium state. Also, note that at this point the perturbed or variational equations are not related to any formal perturbation or variational process, although an intuitive relation can be easily visualized.

2.2 Equilibrium phase plane analysis

We shall, in this section, give an introduction to the analysis of a one-degree--of-freedom system, in the *phase plane*. This plane is the natural result of setting $n = 1$ in our previously defined $2n$-dimensional phase space. In this plane we can investigate Lyapunov stability or orbital stability. In the first instance, then, we are interested in the stability of *points* in the phase plane, i.e., equilibrium points. In the second case, we are interested in the stability of closed curves in the plane. If we limit our discussion to autonomous systems, we have from equations (2.1)

$$\dot{x}_1 = X_1(x_1, x_2), \quad \dot{x}_2 = X_2(x_1, x_2) \tag{2.19}$$

Let us assume that equations (2.19) are variational equations, so that we are interested in investigating here the stability of the origin, $x_1 = x_2 = 0$. If, near the origin, we can cast the X_i in the form

$$\dot{x}_1 = a_{11}x_1 + a_{12}x_2 + \varepsilon_1(x_1, x_2)$$

$$\dot{x}_2 = a_{21}x_1 + a_{22}x_2 + \varepsilon_2(x_1, x_2) \tag{2.20}$$

we can then state the following theorem, [4, p. 187]:

THEOREM 2.1: *If the origin is an isolated equilibrium point and if the nonlinear terms $\varepsilon_i(x_j)$ are such that $\lim_{r \to 0} (\varepsilon_i/r) = 0$, where $r^2 = \| x \|$, the character of the singularity of the linear first-approximation equations $\dot{x}_i = a_{ij}x_j$ is the same as for the complete equations, except where the equilibrium point of the linear equations is a center. Such a point would be a* critical *point, and in that case the equilibrium point of the complete equations is either a center or a focal point.*

Now, we have not yet defined all the terms in Theorem 2.1, e.g., focal point and center. Even so, this theorem has strong implications for our work. For it says, basically, that the linear approximations to equations (2.20), i.e.,

$$\dot{x}_1 = a_{11}x_1 + a_{12}x_2$$

$$\dot{x}_2 = a_{21}x_1 + a_{22}x_2 \tag{2.21}$$

determine – in most cases – the behavior of the nonlinear equations at the equilibrium point. Since the system (2.20) is autonomous, the a_{ij} are constants, and an exponential solution might be sought for equations (2.21). Then, for

$$x = Ae^{\lambda t} \tag{2.22}$$

we obtain an equation for two roots λ_1, λ_2 as

$$\lambda_{1,2} = \frac{P}{2} \pm \sqrt{\left(\frac{P}{2}\right)^2 - Q} \tag{2.23}$$

where

$$P = a_{11} + a_{22}, \quad Q = a_{11}a_{22} - a_{12}a_{21} \tag{2.24}$$

Then, depending on the discriminant of the radical in equation (2.23), we obtain the following cases (see figures 1, 2):

1) $P^2 > 4Q$: Two real distinct roots.

$$\begin{aligned} &Q > 0, P < 0, \quad \textit{stable node} \\ &Q > 0, P > 0, \quad \textit{unstable node} \\ &Q < 0, \qquad\quad \textit{saddle point} \end{aligned}$$

2) $P^2 = 4Q$: Two equal roots.

3) $P^2 < 4Q$: Two complex roots, if $Q > 0$.

$$\begin{aligned} &Q > 0, P < 0, \quad \textit{stable focus} \\ &Q > 0, P > 0, \quad \textit{unstable focus} \\ &Q > 0, P = 0, \quad \textit{center} \end{aligned}$$

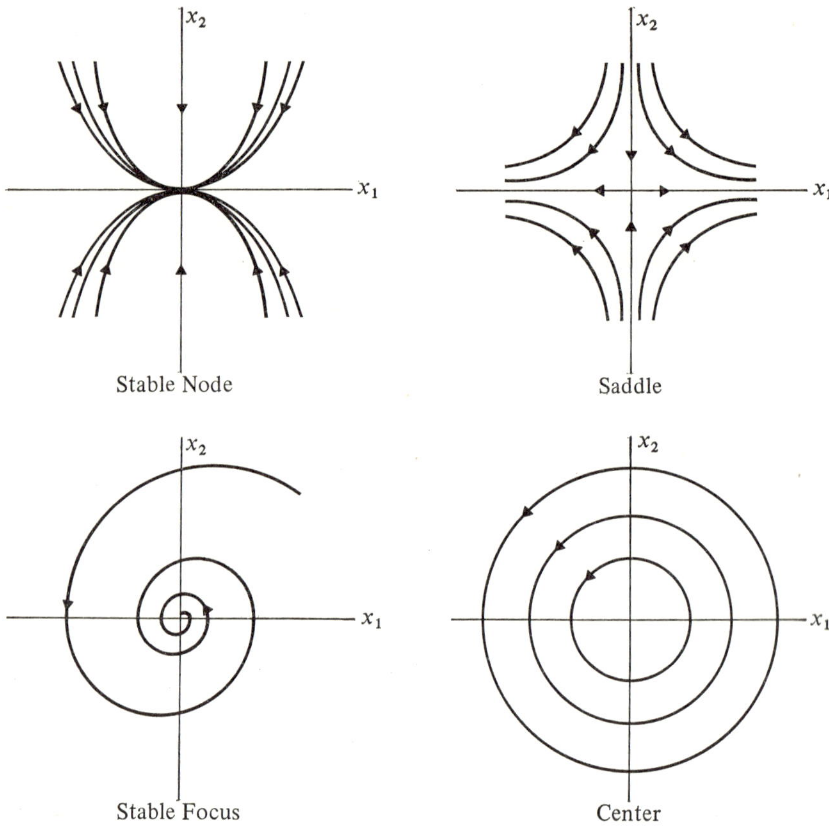

Stable Node

Saddle

Stable Focus

Center

Fig. 1. Equilibrium points in the phase plane.

These results can also be recast in terms of the roots λ_1, λ_2, since from the quadratic formula (2.23) it follows that

$$\lambda_1 + \lambda_2 = P, \ \lambda_1\lambda_2 = Q \tag{2.25}$$

Then we can say that: if λ_1, λ_2 are real and of the same sign, we find either a stable or an unstable *node*; if λ_1, λ_2 are real and of different sign, we obtain a *saddle point*; if λ_1, λ_2 are complex, they must be conjugates, and so we obtain a stable or an unstable *focus* according to the sign of P; and if the roots are purely imaginary we obtain a *center*.

It is of more than casual interest to note that the center, which is here stable in the sense of Lyapunov because we are discussing small perturbations of equilibrium and hence are not restricted to the definition of orbital stability, is the one result which may be in error as we go back from the linear system (2.21) to the complete system (2.20). The center is also the boundary between foci (see figure 2), and hence with any damping at all is in reality likely to be one of the foci.

As an example, let us consider an undamped simple pendulum, with the usual pendulum notation, but with the added proviso that we shall examine stability about both $\theta = 0$ and $\theta = \pi$. For the former, the usual pendulum equation is

$$\ddot{\theta} + \omega^2 \sin \theta = 0, \quad \omega^2 = \frac{g}{L} \tag{2.26}$$

For the inverted pendulum, replace θ by $\theta + \pi$, so that equation (2.26) becomes

$$\ddot{\theta} - \omega^2 \sin \theta = 0 \tag{2.27}$$

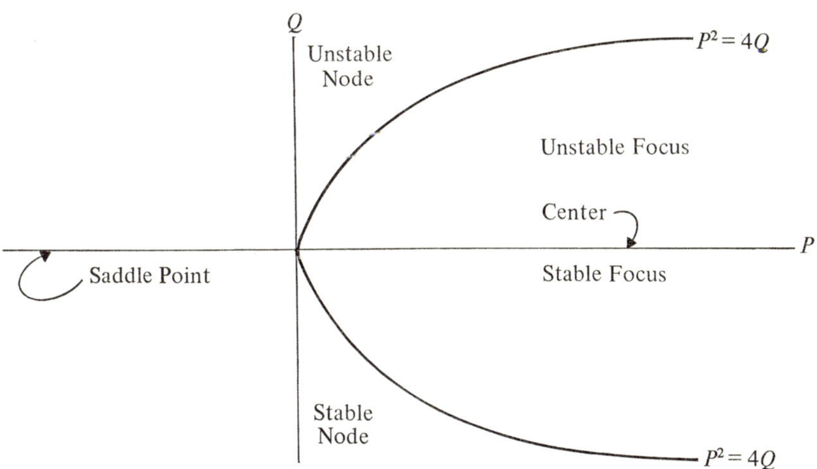

Fig. 2. Characterization of equilibrium point roots, after Meirovitch [4].

13

If we consider now only small perturbations, then we can write the first system in first order form as

$$\dot{x}_1 = x_2, \quad \dot{x}_2 = -\omega^2 x_1 \tag{2.28}$$

while for the inverted pendulum

$$\dot{x}_1 = x_2, \quad \dot{x}_2 = \omega^2 x_1 \tag{2.29}$$

For the normal pendulum system we may easily identify $P = 0$, $Q = \omega^2$, so that we must have two imaginary roots, and hence we have a *center*. According to Theorem 2.1, the stability of the complete pendulum system (2.26) is thus in doubt. Meirovitch [4, pp. 187-189] has shown that if the pendulum is damped, a stable focus is obtained, which is not an unexpected result, for the linear pendulum.

For the inverted pendulum, we can find that $P = 0$, $Q = -\omega^2$, so that the roots λ_1, λ_2 are real but of opposite sign. Thus the inverted pendulum has the behavior of a saddle point, and it must be so for both the linear and nonlinear pendulum equations. The saddle point, of course, is clearly accounted unstable, which readily agrees with our physical intuition.

2.3   General phase plane analysis

In the previous section we have indicated how phase plane analyses would proceed to evaluate the stability of the origin, i.e., of an equilibrium point. This was done in accord with the definition of Lyapunov stability, and it also highlighted the usefulness of the theorem on nonlinearities.

In this section we shall consider the orbital stability of Poincaré in the process of evaluating stability of systems away from equilibrium points. Thus, we will be considering *stability in the large* [4].

Let us consider a one-dimensional conservative system governed by the differential equation

$$\ddot{y} - f(y) = 0 \tag{2.30}$$

Note that the equation (2.30) is autonomous, as $f(y)$ is taken to be an analytic function of y only. In the first order form we have

$$\dot{x}_1 = x_2, \quad \dot{x}_2 = f(x_1) \tag{2.31}$$

If the second of equations (2.31) is multiplied by x_2 we can find

$$x_2 \dot{x}_2 = f(x_1) x_2$$

or

$$\frac{\mathrm{d}}{\mathrm{d}t}(\tfrac{1}{2}x_2^2) = f(x_1)\dot{x}_1 = \frac{\mathrm{d}}{\mathrm{d}t}\int_0^{x_1} f(\eta)\,\mathrm{d}\eta \tag{2.32}$$

If we now define the *potential energy* $U(x_1)$ as

$$U(x_1) = \int_{x_1}^0 f(\eta)\,\mathrm{d}\eta = -\int_0^{x_1} f(\eta)\,\mathrm{d}\eta \tag{2.33}$$

and recognize the *kinetic energy* $T(x_2)$ as

$$T(x_2) = \tfrac{1}{2}x_2^2 \tag{2.34}$$

we see that equation (2.32) reads as

$$\frac{\mathrm{d}}{\mathrm{d}t}(U + T) = \frac{\mathrm{d}E}{\mathrm{d}t} = 0 \tag{2.35}$$

where $E = U + T$ is the total energy. Thus, for E a constant, we have

$$\tfrac{1}{2}x_2^2 + U(x_1) = E \tag{2.36}$$

What is the significance of equation (2.36), particularly for phase plane analysis? First of all that equation represents a *first integral* of the system of the system (2.31), or of the equivalent equation (2.30) for which we might write

$$\tfrac{1}{2}\dot{y}^2 + U(y) = E \tag{2.36a}$$

In the present context of a conservative system, the first integral represents a statement of the principle of *conservation of energy*. In the phase plane, equation (2.36) actually represents a family of curves that depend on the parameter E. These curves are symmetric with respect to the x_1 axis. If an

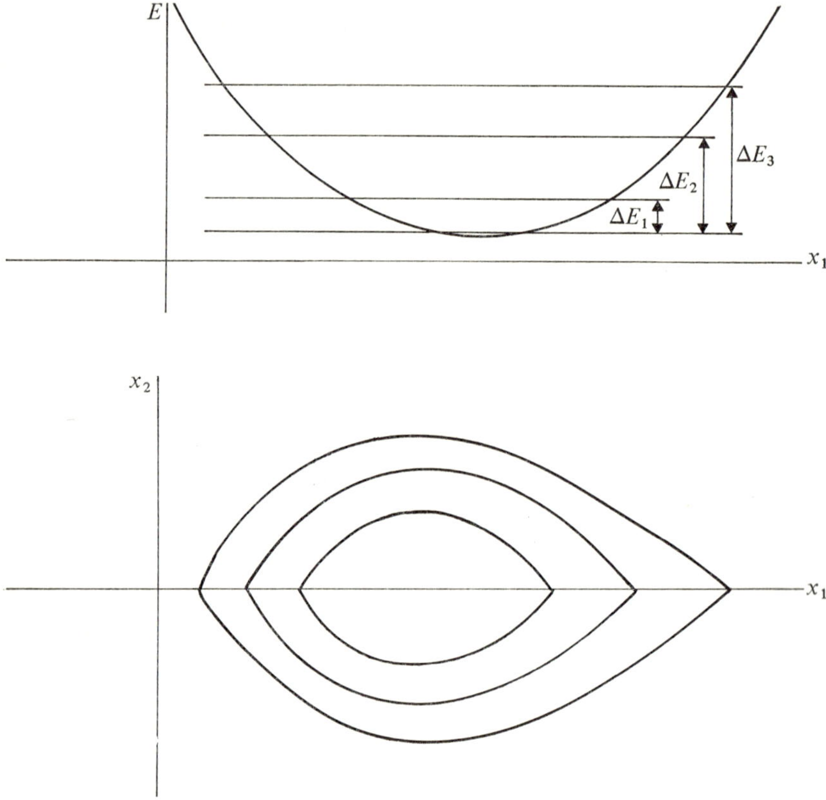

Fig. 3. Level curves in the phase plane.

E axis was constructed normal to the $x_1 x_2$ plane (figure 3), we could envision
equation (2.36) as describing *level curves* resulting from the intersection of
planes $E = \text{const.}$ with surfaces $E = E(x_1, x_2)$. This description allows us
to intuit that the surface takes on extreme values, e.g., a maximum, minimum
or inflection, when $\partial E/\partial x_1$ vanishes. Since $T = T(x_2)$ only, this implies that
the extreme value is characterized by an extremum of the potential energy,
i.e., $\partial U/\partial x_1 = 0$.

Further, it is not difficult to argue that an equilibrium point found by
such a condition would be a center or a saddle point, inasmuch as nodes
and foci require solutions to approach (or move away from) equilibrium as
$t \rightarrow \infty$. The level curve notion requires that the solutions are such that the

integral (2.36) has the same value for any point on a level curve, for all regular points, even surrounding a singular (equilibrium) point. Thus an asymptotic solution, as required for nodes and foci, where the integral curves approach the singular point, is clearly ruled out.

As a simple example, consider the case of a linear spring, with its equilibrium position at $x_1 = a$, and a specific stiffness of k/m. Then

$$f(x_1) = \frac{k}{m}(x_1 - a) \tag{2.37}$$

so that

$$U(x_1) = \frac{k}{2m}[a^2 - (x_1 - a)^2]$$

$$= U(a) - \frac{k}{2m}(x_1 - a)^2 \tag{2.38}$$

At the equilibrium point $x_1 = a$, $x_2 = 0$, we have $U(a) = E(a, 0)$ so that we can write

$$\Delta E = E(x_1, x_2) - E(a, 0) = \frac{1}{2}x_2^2 - \frac{k}{2m}(x_1 - a)^2 \tag{2.39}$$

which is a form that can be used to calculate the energy increments depicted in figure 3.

Further, let us use some elementary results of calculus to examine the equilibrium points of the function of two variables, $E(x_1, x_2)$. In particular, though, note that the construction of the total energy is such (see equations (2.33), (2.34), and (2.36)) that $\partial^2 E/\partial x_1 \partial x_2$ vanishes. Then, for an extreme value we require that

$$\frac{\partial E}{\partial x_1} = \frac{\partial U}{\partial x_1} = 0, \qquad \frac{\partial E}{\partial x_2} = \frac{\partial T}{\partial x_2} = 0 \tag{2.40}$$

while the nature of the extreme value is determined by the following possibilities. If Δ is defined as

$$\Delta = \frac{\partial^2 E}{\partial x_1^2}\frac{\partial^2 E}{\partial x_2^2} - \left(\frac{\partial^2 E}{\partial x_1 \partial x_2}\right)^2 = \frac{\partial^2 U}{\partial x_1^2}\frac{\partial^2 T}{\partial x_2^2} = \frac{\partial^2 U}{\partial x_1^2} \tag{2.41}$$

then the energy will take on a minimum point if $\partial^2 U/\partial x_1^2 > 0$ at the equilibrium point, and it will be a saddle point if $\partial^2 U/\partial x_1^2 < 0$ at the equilibrium point. These details follow in this way because the kinetic energy is such that $\partial T/\partial x_2 = 0$ and $\partial^2 T/\partial x_2^2 > 0$ at equilibrium.

Thus it follows that the nature of the energy extremum can be determined by examination of the potential energy only, and in fact we could posit that *if the potential energy $U(x_1)$ has a minimum at an equilibrium point, the equilibrium is stable.* Further, we could extrapolate also that *if the potential energy is not a minimum, then the equilibrium will not be stable.* The first of these ideas will be formalized as Dirichlet's theorem in the sequel, while the second is credited to Lyapunov, and it too will be formalized in the sequel.

For the linear spring considered earlier, note that

$$\left.\frac{\partial U}{\partial x_1}\right|_{x_1=a} = 0, \qquad \left.\frac{\partial^2 U}{\partial x_1^2}\right|_{x_1=a} = -\frac{k}{m} \tag{2.42}$$

so that the stability depends on whether the spring is a restoring force or not. Looking back at equations (2.30) and (2.37), we see that $k < 0$ is required to get the conventional spring-mass system equation to correspond to a restoring force in the spring. Not surprisingly, then, we see that $k < 0$ implies that $\partial^2 U/\partial x_1^2 > 0$ and so the equilibrium is stable.

To close this chapter, we should like to briefly discuss the concept of Poincaré's *limit cycles*. We have just seen that, for the linear spring-mass system, the equilibrium is stable, and it clearly is of the center type. That is, if the equilibrium point is at the origin ($a = 0$) and the spring is of a restoring type, then

$$\dot{x}_1 = x_2, \quad \dot{x}_2 = -\omega^2 x_1, \quad \omega^2 = \frac{k}{m}$$

$$E = \frac{1}{2}x_2^2 + \frac{1}{2}\omega^2 x_1^2 = \text{const.} \tag{2.43}$$

Recall that in our discussion of the pendulum, the differential equations (2.43) were shown to have a center as an equilibrium point. Now the solutions of these equations are periodic, and thus the closed curves in the phase plane represent periodic motion about an equilibrium point.

It also happens that for nonconservative nonlinear systems, the phenomenon of cyclic trajectories appears frequently. Consider for example the following nonlinear system:

$$\dot{x}_1 = x_2 + x_1(1 - \sqrt{x_1^2 + x_2^2})\,(2 - \sqrt{x_1^2 + x_2^2})$$

$$\dot{x}_2 = -x_1 + x_2(1 - \sqrt{x_1^2 + x_2^2})\,(2 - \sqrt{x_1^2 + x_2^2}) \qquad (2.44)$$

If we examined the linearized system

$$\dot{x}_1 = x_2 + 2x_1, \quad \dot{x}_2 = -x_1 + 2x_2 \qquad (2.45)$$

we could readily verify that the origin – the equilibrium point – is an unstable focus, with characteristic roots $\lambda_{1,2} = 2 \pm i$. However, in order to illustrate the limit cycle phenomenon, we shall look for stability in the large, away from the equilibrium point. That is, we shall now look at where the trajectories that are unstable at the origin lead to over the remainder of the phase plane.

It is convenient to introduce polar coordinates in the plane, i.e.,

$$x_1 = \rho \cos \theta, \quad x_2 = \rho \sin \theta \qquad (2.46)$$

so that the differential equations (2.44) become

$$\dot{\rho} \cos \theta - \rho \sin \theta \dot{\theta} = \rho \sin \theta + \rho \cos \theta\,(1 - \rho)\,(2 - \rho)$$

$$\dot{\rho} \sin \theta + \rho \cos \theta \dot{\theta} = -\rho \cos \theta + \rho \sin \theta\,(1 - \rho)\,(2 - \rho) \qquad (2.47)$$

from which we may obtain

$$\dot{\rho} = \rho(1 - \rho)\,(2 - \rho), \quad \dot{\theta} = -1 \qquad (2.48)$$

Clearly the distance from the origin to any point on the phase plane is given by ρ. Also, we see that $\dot{\rho} = 0$ at the origin, and at $\rho = 1$, $\rho = 2$. Also, note that

$$\dot{\rho} > 0 \text{ for } 0 < \rho < 1$$

$$\dot{\rho} < 0 \text{ for } 1 < \rho < 2 \qquad (2.49)$$

$$\dot{\rho} > 0 \text{ for } 2 < \rho$$

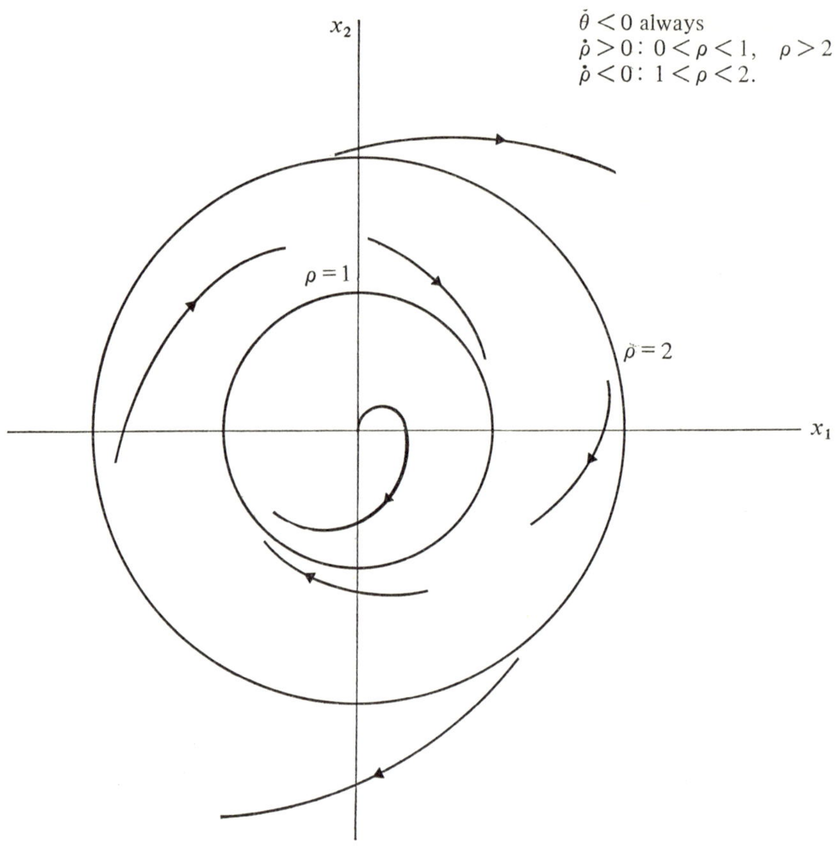

Fig. 4. Phase plane plots for equations (2.44, 2.48).

Thus we can identify two closed curves in the phase plane, corresponding to $\rho = 1$ and $\rho = 2$, which are *limit cycles*. Solutions emanating from the origin and from the annulus $1 < \rho < 2$ all tend to the *stable limit cycle* $\rho = 1$, as $t \to \infty$. The curve $\rho = 2$ is an *unstable limit cycle* because for $\rho > 2, \dot\rho > 0$, and for $\rho < 2, \dot\rho < 0$, so that the trajectories move away from the curve $\rho = 2$, either towards infinity or towards the stable limit cycle $\rho = 1$. These results are pictured in figure 4.

2.4   Summary

We have presented in this chapter some definitions of stability, some notions of variational equations and of linearized equations, and a discussion of the types of equilibrium possible at the singular points of linear (or linearized) systems. This last is particularly useful in view of the important Theorem 2.1 that relates the equilibrium behavior of nonlinear systems to that of the corresponding linearized equations.

We have also discussed stability in the large in the phase plane, and introduced the notion of level curves and of first integrals. We used an example of a conservative system to get at a preliminary discussion of the theorems of Dirichlet and of Lyapunov. The chapter was concluded with a study of the stability in the large of a nonlinear system. This served to demonstrate the existence and meaning of Poincaré limit cycles, as well as to relate equilibrium investigations to stability in the entire phase plane.

References

[1]   F. G. Tricomi: *Differential Equations.* Hafner Publishing Company, New York, 1961.

[2]   W. Hahn: *Stability of Motion.* Springer-Verlag, Berlin, 1967.

[3]   J. LaSalle and S. Lefschetz: *Stability by Liapunov's Direct Method.* Academic Press, New York, 1961.

[4]   L. Meirovitch: *Methods of Analytical Dynamics.* McGraw-Hill, New York, 1970.

[5]   H. Leipholz: *Stability Theory.* Academic Press, New York, 1970.

3

Autonomous system stability

3.1 Criteria for linear autonomous systems

In Chapter 2 we discussed the stability at equilibrium points of single-degree-
-of-freedom systems, particularly linear, autonomous systems. Those results
were particularly simple because it required only the solution of a quadratic
equation to determine the nature of the characteristic roots. Obviously a
multi-degree-of-freedom system will be more complex, and it will be examined
now in a number of different ways.

Consider the linear (or linearized) autonomous system

$$\dot{x}_i = a_{ij}x_j, \quad i,j = 1 \ldots n \tag{3.1}$$

or

$$\dot{x} = [a_{ij}]x = Ax \tag{3.2}$$

which has an equilibrium point at $x = \dot{x} = 0$. For a linear system, the standard solution form is $\exp(\lambda t)$, leading to the characteristic equation

$$|\, a_{ij} - \lambda \delta_{ij}\,| = 0, \quad A - \lambda I = 0 \tag{3.3}$$

or its equivalent polynomial

$$f(\lambda) = \lambda^n + a_1 \lambda^{n-1} + \ldots + a_{n-1}\lambda + a_n = 0 \tag{3.4}$$

Stability at the equilibrium point is clearly dependent on the roots of the
polynomial (3.4), and so we can say the following:

1)   If all the roots λ_i have negative real parts, the origin is asymptotically stable.

2)   If there is a single root with a positive real part, the origin is unstable.

3)   If there are roots whose real part is zero, the stability of the origin depends on the multiplicity of these roots. One such root permits stability, although a pair of such (equal) roots does not, for a solution type $(A + Bt)$ is required. Further, for just a single such root, the stability obviously cannot by asymptotic. As a simple example, consider the system

$$\dot{x}_1 = -3x_1$$

$$\dot{x}_2 = -3x_2$$

$$\dot{x}_3 = -3x_1 + x_4 - x_5 \qquad (3.5)$$

$$\dot{x}_4 = 6x_2$$

$$\dot{x}_5 = -x_5$$

which has the stability (frequency) determinant

$$\begin{vmatrix} \lambda + 3 & 0 & 0 & 0 & 0 \\ 0 & \lambda + 3 & 0 & 0 & 0 \\ 3 & 0 & \lambda & -1 & 1 \\ 0 & -6 & 0 & \lambda & 0 \\ 0 & 0 & 0 & 0 & \lambda + 1 \end{vmatrix} = 0 \qquad (3.6)$$

which in turn can be expanded by minors to yield

$$\lambda^2 (\lambda + 3)^2 (\lambda + 1) = 0 \qquad (3.7)$$

Thus the roots are

$$\lambda = 0, 0, -1, -3, -3 \qquad (3.8)$$

so that the origin must be unstable. Remember, the multiplicity of a root such as $\lambda = -3$ is not a problem, since $(A + Bt) \exp(-3t)$ still vanishes as $t \to \infty$. However, this is not so for $\lambda = 0$, and so the origin is unstable.

In the above example, because of the simple nature of equations (3.5), there were many zero entries in the determinant (3.6) and so its expansion was not difficult. Further, this simple example led to an easily factored characteristic equation. The non-trivial nature of the expansion of most frequency determinants suggests that the characteristic equation of an *n*th order differential equation should be obtained prior to transforming it into a system of the form (3.2).

Insofar as dealing with the characteristic polynomial itself, we will present here two methods of ascertaining whether or not the roots contain negative real parts. These methods are called the stability criteria of *Nyquist* and of *Routh and Hurwitz* [1, 2]. The criteria are phrased in terms of an *n*th order polynomial

$$f(\lambda) = \lambda^n + a_1\lambda^{n-1} + \ldots + a_n \tag{3.9}$$

which is called a *Hurwitz polynomial* if the roots λ_v are such that $R_e(\lambda_v) < 0$ for $v = 1, 2 \ldots n$.

In figure 1 we have displayed two complex planes, i.e., the plane of the variable $\lambda = x + iy$ and the plane of $f(\lambda) = u + iv$. In the former plane, the roots λ_v are clustered in the left half-plane, if $f(\lambda)$ is a Hurwitz polynomial. The roots λ_v must then correspond to the origin $f(\lambda_v) = 0$ in the $f(\lambda)$ plane.

With these two diagrams we can lay out the *Nyquist criterion*. In the left plane we notice that as we traverse the *iy* axis from $y = -\infty$ to $y = +\infty$, all the roots λ_v stay to our left. *Then it follows that if we were to graph $f(iy)$ for the same increasing values of y, then as we follow the curve in the $f(\lambda)$*

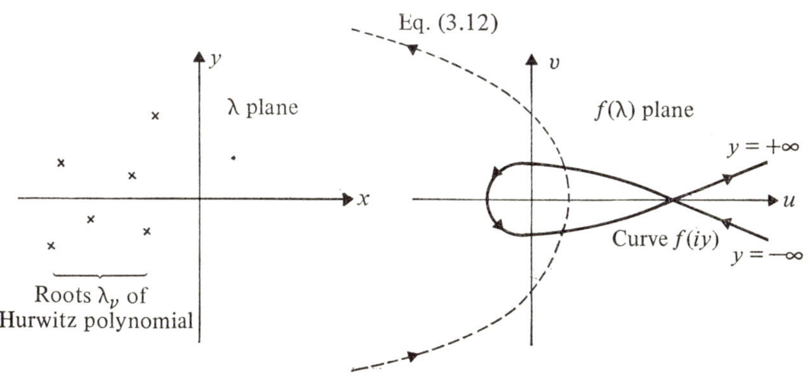

Fig. 1. Diagrams for the Nyquist criterion.

plane the origin $f(\lambda_v) = 0$ should always be on our left, for a stable origin.
As a simple example, consider the polynomial

$$f(\lambda) = \lambda^2 + \lambda + \tfrac{1}{2} \tag{3.10}$$

It is easy to show that (3.10) is a Hurwitz polynomial, with roots

$$\lambda_v = -\frac{1}{2} \pm \frac{i}{2} \tag{3.11}$$

In the $f(\lambda)$ plane, we can write $f(iy)$ as

$$f(iy) = iy + \tfrac{1}{2} - y^2 \tag{3.12a}$$

or

$$u = \tfrac{1}{2} - y^2, \quad v = y \tag{3.12b}$$

In the u, v plane, equation (3.12b) is a parabola, as indicated in figure 1, with increasing y as shown.
On the other hand, for the polynomial

$$f(\lambda) = \lambda^2 - \lambda + \tfrac{1}{2} \tag{3.13}$$

the roots are

$$\lambda_v = \frac{1}{2} \pm \frac{i}{2} \tag{3.14}$$

corresponding to unstable behavior. It is easy enough to demonstrate that for the polynomial (3.13), $f(iy)$ is such that

$$u = \tfrac{1}{2} - y^2, \quad v = -y \tag{3.15}$$

so that the curve is as shown in figure 1, but directed the other way as y increases! Thus, for the unstable case, the polynomial (3.13) is not a Hurwitz polynomial, and the origin is on the right side of the curve $f(iy)$ as y increases.
To introduce the Routh-Hurwitz criterion, let us examine the following Hurwitz polynomial, of degree three,

$$f(\lambda) = \lambda^3 + a_1\lambda^2 + a_2\lambda + a_3 \tag{3.16}$$

There are two possible sets of root constructions for a cubic Hurwitz polynomial, i.e.,

$$\lambda_i = -\sigma_i, \quad i = 1, 2, 3, \quad \sigma_i > 0 \tag{3.17a}$$

$$\lambda_1 = -\sigma_1, \quad \lambda_{2,3} = -\sigma_2 \pm i\omega^2, \quad \sigma_i > 0 \tag{3.17b}$$

With these roots, the corresponding polynomials can be written as

$$\lambda^3 + (\sigma_1 + \sigma_2 + \sigma_3)\lambda^2 + (\sigma_1\sigma_2 + \sigma_1\sigma_3 + \sigma_2\sigma_3)\lambda + \sigma_1\sigma_2\sigma_3 = 0 \tag{3.18a}$$

$$\lambda^3 + (\sigma_1 + 2\sigma_2)\lambda^2 + (2\sigma_1\sigma_2 + \sigma_2^2 + \omega^2)\lambda + \sigma_1(\sigma_2^2 + \omega^2) = 0 \tag{3.18b}$$

Then by comparing equations (3.18a, b) with equation (3.16), we observe first of all that

$$a_j > 0, \quad j = 1, 2, 3 \tag{3.19}$$

and further, that

$$a_1 a_2 - a_3 > 0 \tag{3.20}$$

Thus, for the cubic polynomial (3.16), we have inferred some of its properties on the basis of assumed stability behavior as reflected in the nature of the polynomial's roots.

The generalization of these features revolves around the *Hurwitz matrix* H, which is defined for the nth order polynomial (3.9) as [2, 3]

$$H = \begin{bmatrix} a_1 & 1 & 0 & 0 \ldots\ldots 0 \\ a_3 & a_2 & a_1 & 1 \ldots\ldots 0 \\ a_5 & a_4 & a_3 & a_2 \ldots\ldots \\ \cdots\cdots\cdots\cdots\cdots\cdots\cdots \\ 0 & 0 & 0 & \ldots\ldots a_n \end{bmatrix} \tag{3.21}$$

where the diagonal is constructed of the coefficients $a_1, a_2 \ldots a_n$, and the first column of the elements $a_1, a_3 \ldots a_n$ or a_{n-1} (depending on whether n is odd or even) and thereafter of zeroes. Thus, in fact, we have for H an nth order array in which the bottom row contains only zeroes but for the last element on the diagonal. Then we may state the following theorem:

THEOREM 3.1 (Routh-Hurwitz): *A necessary and sufficient condition for the polynomial (3.9) to be a Hurwitz polynomial is that all of the principal minors* $\Delta_1, \Delta_2 \ldots \Delta_n$ *of the Hurwitz matrix H be positive.*

Thus, for stability, the theorem requires that

$$\Delta_1 = a_1 > 0$$

$$\Delta_2 = \begin{vmatrix} a_1 & 1 \\ a_3 & a_2 \end{vmatrix} > 0$$

$$\Delta_3 = \begin{vmatrix} a_1 & 1 & 0 \\ a_3 & a_2 & a_1 \\ a_5 & a_4 & a_3 \end{vmatrix} > 0 \tag{3.22}$$

down to the last principal minor, whose positivity requirement can be written as

$$\Delta_n = a_n \Delta_{n-1} \tag{3.23}$$

because of the aforementioned properties of the last row.

As an example we shall consider the damped pendulum, in both the standard and the inverted configurations. The general equation can be written as

$$\dot{x}_1 = x_2, \quad \dot{x}_2 = -(g/L)\sin x_1 - (c/m)x_2 \tag{3.24}$$

where c is the damping coefficient and $x_1 = \theta$ is the pendulum swing angle. For infinitesmal motion in the routine configuration, about $x_1 = x_2 = 0$, we examine the system

$$\dot{x}_1 = x_2, \quad \dot{x}_2 = -(g/L)x_1 - (c/m)x_2 \tag{3.25}$$

which has as a characteristic polynomial

$$\lambda^2 + (c/m)\lambda + g/L = 0 \tag{3.26}$$

The principal minors of the polynomial (3.26) are easily found to be

$$\Delta_1 = c/m > 0, \quad \Delta_2 = (g/L)\Delta_1 > 0 \tag{3.27}$$

Thus the standard pendulum is always stable if the damping is dissipative, i.e., if $c > 0$.

For the inverted pendulum the equilibrium point is $x_1 = \pi$, $x_2 = 0$, so that as in Chapter 2 we consider here the perturbed system

$$\dot{x}_1 = x_2, \quad \dot{x}_2 = (g/L)\, x_1 - (c/m)\, x_2 \tag{3.28}$$

whose characteristic polynomial is

$$\lambda^2 + (c/m)\, \lambda - g/L = 0 \tag{3.29}$$

Here the principal minors are

$$\varDelta_1 = c/m > 0, \quad \varDelta_2 = (- g/L)\, \varDelta_1 < 0 \tag{3.30}$$

so that the inverted pendulum is always unstable, irrespective of the type of damping in the system.

We note in closing this part of our discussion that we have presented two methods of examining the roots of a polynomial for negative real parts. One method has been geometric, or graphic, and the other analytic. In both instances the value lies in the fact that the polynomial itself does not have to be factored or otherwise solved. All that is required is some manipulation of the polynomial coefficients.

However, we must remember that these two criteria, powerful as they might be, are restricted to linear systems, in particular linear systems near an equilibrium point. While Theorem 2.1 (see Section 2.2) served to highlight the importance of linear system investigations for local stability, the example at the close of Section 2.3 indicated a need for developing some methodology both for dealing with global stability and with nonlinearity. The next section begins such an attack.

3.2 The Lyapunov function—Definition and stability theorems

We introduce now *Lyapunov's direct (second) method*, the idea being to use directly the differential equations that govern the system behavior, without actually solving them, and without recourse to the variational equations. As there is no universally applicable method of producing Lyapunov functions, this approach could be termed in some sense philosophical; nonetheless it is both important and useful. The method is not restricted to linear systems, and may be used to examine stability in the large. It was inspired by Dirichlet's proof of Lagrange's theorem on the stability of equilibrium of a system, and thus there is (for conservative systems) a strong relationship between Lyapunov functions and energy (first integral) functions.

We emphasize that Lyapunov functions are not unique. Also, if one cannot be produced for a given problem, it does not imply that the system being examined is not stable. It simply means that a Lyapunov function has not been found!!

We will confine ourselves to discrete systems governed by differential equations of the form

$$\dot{x} = X(x) \tag{3.31}$$

The relationship $X(0) = 0$ is satisfied, so that the origin is a singular (stationary, equilibrium) point. For any initial vector $x(t_0) = x_0$, a unique solution to the system (3.31) exists.

Consider a real continuous function $V(x)$ with appropriate continuity properties and differentiability properties, with $V(0) = 0$, and the following definitions:

1) $V(x)$ is *positive (negative) definite* if $V(x) > 0 \, (< 0)$ for all $x \neq 0$, and $V(0) = 0$.

2) $V(x)$ is *positive (negative) semi-definite* if $V(x) \geq 0 \, (\leq 0)$ and it can vanish also for some $x \neq 0$.

3) $V(x)$ is *indefinite* if it can assume both positive and negative values in the domain of interest.

We may now compute the total derivative of the function $V(x)$ along a trajectory of the system (3.31) as

$$\dot{V}(x) = \frac{\mathrm{d}V(x)}{\mathrm{d}t} = \frac{\partial V(x)}{\partial x_i} \frac{\mathrm{d}x_i}{\mathrm{d}t} = \mathrm{grad} \, V \cdot \dot{x} \tag{3.32}$$

where the x_i are components of vector x and grad $V = \nabla V$ is the gradient of the scalar $V(x)$. The calculation of $\dot{V}$ is the mechanism that brings the differential equation system into the problem, without our knowing its solution!

We will now state a series of stability and instability theorems [1-7], without proof. It may be noted that extended versions of these may also be stated, with relaxed requirements. However, as these require greater mathematical precision, and as the former are adequate for our purposes, we will not state the extensions here.

The first two theorems are called the *Lyapunov stability theorems:*

THEOREM 3.2: *If there exists for the system* (3.31) *a positive (negative) definite* $V(x)$ *whose total time derivative* $\dot{V}(x)$ *is negative (positive) semi-definite along every trajectory of* (3.31), *then the origin is Lyapunov stable.*

THEOREM 3.3: *If there exists for the system* (3.31) *a positive (negative) definite function* $V(x)$ *whose total time derivative* $\dot{V}(x)$ *is negative (positive) definite along every trajectory of* (3.31), *then the trivial solution is asymptotically Lyapunov stable.*

(Note: the statement 'the trivial solution $x = 0$ is stable' is equivalent to our earlier statements that 'the origin is stable'.)

Thus the first theorem allows for centers and for orbital stability; the latter indicates asymptotic stability, a stronger condition!

The next two theorems are termed *Lyapunov instability theorems*:

THEOREM 3.4: *If there exists for the system* (3.31) *a function* $V(x)$ *whose total time derivative* $\dot{V}(x)$ *is positive (negative) definite along every trajectory of* (3.31), *and if the function itself can assume positive (negative) values for arbitrarily small values of* x, *then the trivial solution is unstable.*

Note that this theorem requires that $\dot{V}(x)$ be the same sign as $V(x)$, and that $\dot{V}(x)$ be definite, in a small spherical neighborhood ($\| x \| < \varepsilon$) of the origin. We shall loosen this restriction in Theorem 3.6.

THEOREM 3.5: *If there exists for the system* (3.31) *a function* $V(x)$ *such that*

$$\dot{V}(x) = \lambda V(x) + W(x)$$

along every trajectory of (3.31), *where* λ *is a positive constant and* $W(x)$ *is either identically zero or is a positive (negative) function, and if in the latter case the function* $V(x)$ *is not negative (positive), then the trivial solution is unstable.*

The point of these two theorems is that if $V(x)$ and $\dot{V}(x)$ are of the same sign, the system is unstable at $x = 0$ (the equilibrium point).

We now state a somewhat less restrictive instability theorem due to *Chetayev*:

THEOREM 3.6: *If there exists for the system (3.31) a function $V(x)$ such that*
1)   in any arbitrarily small neighborhood of the origin there is a region D_1
in which $V(x) > 0$ and on whose boundaries $V(x) = 0$;
2)   at all points of the region in which $V(x) > 0$ the total time derivative
$\dot{V}(x)$ assumes positive values along every trajectory of (3.31); and
3)   the origin is a boundary point of D_1;
then the trivial solution is unstable.

Thus a complete sphere (circle) around the origin for which V and $\dot{V}$ both
are of the same sign does not have to be produced.

To supply a motivation for these theorems, we will give a geometric
interpretation for a second order system in the phase plane [3]. Consider
first a plot of the surface $z = V(x_1, x_2)$ as shown (figure 2) for a positive
definite $V(x)$. Note, then, that there is a cup tangent to the origin.

Now if $\dot{V}(x_1, x_2)$ is negative definite, then the trajectory must perforce
be moving down the surface $V(x_1, x_2)$ towards the origin. Clearly this means
that in the phase plane the closed curves $V(x_1, x_2) = c_i$ are being intersected

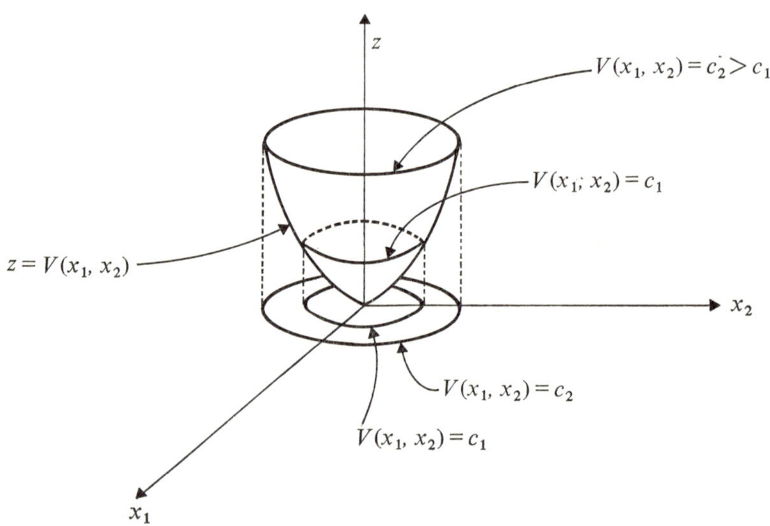

Fig. 2. Schematic diagram of a Lyapunov function, after Meirovitch [3].

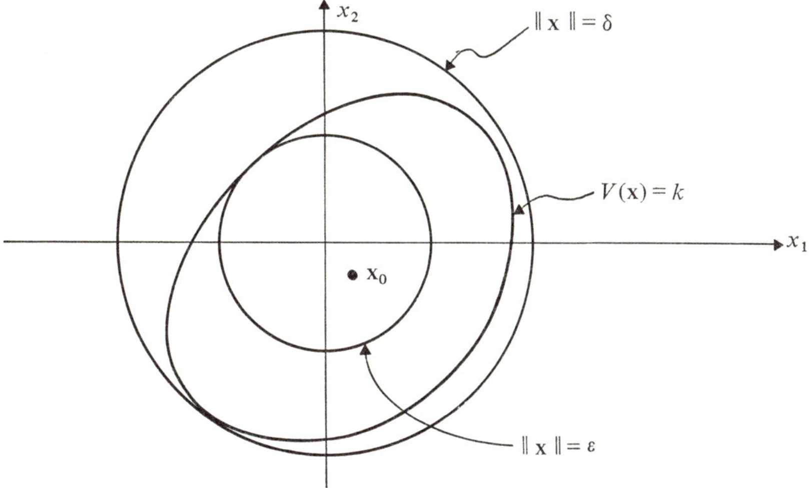

Fig. 3. Stability definitions in the phase plane.

as the trajectory moves towards the origin (asymptotic stability). On the other hand if $\dot{V}(x_1, x_2)$ is positive definite, as is $V(x_1, x_2)$, then the trajectory is climbing the surface (climbing the walls?), and in the phase plane is moving away from the origin, implying instability. If $\dot{V}(x_1, x_2)$ is semi-definite, then the trajectory could 'get stuck' on a particular value of $V(x_1, x_2) = c_i$, which implies stability, although not asymptotic stability.

In figure 3 we have displayed a planar case to indicate how the proofs of he stability theorems might proceed.

For a given constant k, we have depicted a curve in x_1, x_2 space given by $V(x_1, x_2) = k$, together with a circle of radius δ which just touches $V = k$ from the inside, and a circle of radius ε which just touches $V = k$ at the outside. Thus the area inside $V = k$ completely encloses the circle $\| x \| = \varepsilon$, and is in turn enclosed by the circle $\| x \| = \delta$. Now for any intitial point $x(t_0) = x_0$ within $\| x \| = \varepsilon$ it is clear that $V(x_0) < k$, simply because $V(x) = c_i$ represents a set of curves that sequentially enclose each other with increasing values of c, i.e., if $c_1 < c_2$, $V(x) = c_1$ is contained within $V(x) = c_2$. Further, by the first stability theorem (Theorem 3.2), $V(x)$ is at least a non-increasing function along any trajectories emanating from x_0, so that the curve $V(x) = k$ is never reached. Thus any path starting within $\| x \| < \varepsilon$, by appropriate choice of initial conditions, will stay there, and we have Lyapunov stability.

If $\dot{V}(x)$ is clearly negative definite, and not just semi-definite, it follows that a trajectory emanating from x_0 must tend towards the origin, and so toward asymptotic stability. This is true, again, because the curves $V(x) = c_i$ form a nested set of curves with increasing c_i, with the 'smallest'' one being the origin for which $V(0) = 0$. Thus for $V < 0$ the trajectory penetrates the nested set given by $c_n, c_{n-1} \ldots c_2, c_1$, until the origin is reached.

The instability theorems are based on the idea that if $\dot{V}$ is positive the trajectory emanating from x_0 must continue to penetrate curves of increasing values of $V(x) = c_i$, until the curve $\| x \| = \delta$ itself is penetrated. Here, unlike the stability proof, there is no set of initial conditions — or no way to choose ε and $\delta(\varepsilon)$ — to keep the trajectories within a small neighborhood of the origin.

3.3   Some Lyapunov function applications

We present in this section a series of examples of discrete systems to illustrate application of the Lyapunov direct (second) method [3, 4, 2].

EXAMPLE 3.3.1: Consider first the rotation about the center of mass of a torque-free body, with moments of inertia A, B, C about the principal axes 1, 2, 3. Then, from Euler's equations,

$$A\dot{\omega}_1 + (C - B)\omega_2\omega_3 = 0$$

$$B\dot{\omega}_2 + (A - C)\omega_1\omega_3 = 0 \tag{3.33}$$

$$C\dot{\omega}_3 + (B - A)\omega_1\omega_2 = 0$$

The unperturbed motion is given by (for motion about axis 3)

$$\omega_1 = \omega_2 = 0, \quad \omega_3 = \omega_0 = \text{const.} \tag{3.34}$$

We will examine small perturbations of the unperturbed motion, i.e.,

$$\omega_1 = \varepsilon_1, \quad \omega_2 = \varepsilon_2, \quad \omega_3 = \omega_0 + \varepsilon_3 \tag{3.35}$$

Substitution of equations (3.34, 3.35) into equation (3.33) yields

$$A\dot{\varepsilon}_1 + (C - B)\varepsilon_2(\omega_0 + \varepsilon_3) = 0$$

$$B\dot{\varepsilon}_2 + (A - C)\varepsilon_1(\omega_0 + \varepsilon_3) = 0 \tag{3.36}$$

$$C\dot{\varepsilon}_3 + (B - A)\varepsilon_1\varepsilon_2 = 0$$

If we initially examine the *linearized* problem,

$$A\dot{\varepsilon}_1 + \omega_0(C - B)\varepsilon_2 = 0$$

$$B\dot{\varepsilon}_2 + \omega_0(A - C)\varepsilon_1 = 0 \tag{3.37}$$

$$C\dot{\varepsilon}_3 = 0$$

it is easy enough to find for equations (3.37) the following characteristic roots

$$\lambda = 0 \quad \text{and} \quad \lambda = \pm \frac{i\omega_0}{\sqrt{AB}}\sqrt{(C - A)(C - B)} \tag{3.38}$$

If $(C - A)(C - B) < 0$, the system is clearly unstable. If the reverse is the case, we will have a center, and so (by Theorem 2.1) we must examine the nonlinear case.

The nonlinear problem will be examined by developing a series of Lyapunov functions. First, if we multiply the first of equations (3.36) by $(A - C)\varepsilon_1$ and the second by $(C - B)\varepsilon_2$ and then take the difference, we are led to the function

$$V_1 = A(A - C)\varepsilon_1^2 - B(C - B)\varepsilon_2^2 \tag{3.39}$$

By differentiation and substitution one can easily verify that

$$\dot{V}_1 = 0 \tag{3.40}$$

Thus, we can see that as long as $A > C$ and $B > C$, V will be positive definite, and so rotation about the *minimum* axis is stable, by Theorem 3.2.

In a similar way we can construct a Lyapunov function for rotation about the *maximum* axis, i.e., let

$$V_2 = B(C - B)\varepsilon_2^2 - A(A - C)\varepsilon_1^2 \tag{3.41}$$

Again we could easily demonstrate that $\dot{V}_2 = 0$, and that V_2 is positive definite for $C > B$, $C > A$, so that when axis 3 is the maximum axis stability is insured by Theorem 2.1.

Recall that in the examination of the linearized equations, we had determined that the condition $(C - A)(C - B) > 0$ yielded a center at the equilibrium point. This condition is satisfied when C is either the maximum inertia axis or the minimum axis. The fact that the analysis of the nonlinear system yielded a stable result, in the sense of Lyapunov stability, provides an example that demonstrates Theorem 2.1.

Now if we consider the function

$$V_3 = \varepsilon_1 \varepsilon_2 \tag{3.42}$$

then we can show that

$$\dot{V}_3 = (\omega_0 + \varepsilon_3)\left(\frac{C - A}{B}\,\varepsilon_1^2 + \frac{B - C}{A}\,\varepsilon_2^2\right) \tag{3.43}$$

If ε_1 and ε_2 are both positive, and if the rotation axis has the intermediate moment of inertia ($B > C > A$), then both V_3 and $\dot{V}_3$ are positive. By invoking Chetayev's theorem (Theorem 3.6) we can then state that rotation about the intermediate axis is unstable.

EXAMPLE 3.3.2: Consider the example pictured in figure 4, that is, the

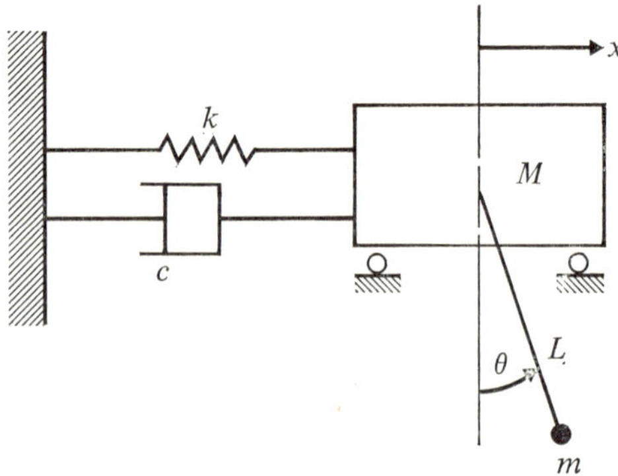

Fig. 4. A coupled, damped, spring-mass-pendulum system.

stability of a pendulum on the moving block. For reference purposes, the kinetic, potential and dissipation functions are written as

$$T = \tfrac{1}{2}\left[M\dot{x}^2 + m(\dot{x} + L\dot{\theta})^2\right]$$

$$V = \tfrac{1}{2}kx^2 + \tfrac{1}{2}mgL\theta^2 \tag{3.44}$$

$$F = \tfrac{1}{2}c\dot{x}^2$$

where, of course, we have assumed $\sin\theta \simeq \theta$. The corresponding equations of motion are:

$$(M + m)\ddot{x} + c\dot{x} + kx + mL\ddot{\theta} = 0$$

$$mL\ddot{x} + mL^2\ddot{\theta} + mgL\theta = 0 \tag{3.45}$$

Choose as a Lyapunov function the *Hamiltonian H* which is the sum of kinetic and potential energies:

$$H = T + V \tag{3.46}$$

which, by the definitions (3.44), is clearly positive definite. Then compute $\dot{H}$, the time derivative:

$$\dot{H} = \dot{T} + \dot{V}$$

$$= M\dot{x}\ddot{x} + m(\dot{x} + L\dot{\theta})(\ddot{x} + L\ddot{\theta}) + kx\dot{x} + mgL\theta\dot{\theta}$$

$$= (M\ddot{x} + m\ddot{x} + mL\ddot{\theta} + kx)\dot{x} + (mL\ddot{x} + mL^2\ddot{\theta} + mgL\theta)\dot{\theta} \tag{3.47}$$

The coefficient of the $\dot{\theta}$ term vanishes by the second of equations (3.44) while the coefficient of $\dot{x}$ is $(-c\dot{x})$ by the first of those equations. Thus it follows that

$$\dot{H} = -c\dot{x}^2 = -2F \tag{3.48}$$

Clearly we will always have Lyapunov stability for the system (3.45), from the Lyapunov function and its derivative given by equations (3.46) and (3.48). When the damping is non-trivial, then $\dot{H} < 0$ always, and we have asymptotic stability. We will show in the next section that in a discrete mechanical system with *pervasive damping*, asymptotic stability is always the resulting behavior.

EXAMPLE 3.3.3: As a simple mathematical example, consider the very famous van der Pol equation:

$$\ddot{y} + y - \varepsilon \left(\frac{\dot{y}^3}{3} - \dot{y} \right) = 0 \tag{3.49}$$

Let

$$x_1 = y, \quad \dot{x}_1 = x_2$$

so that the corresponding system is

$$\dot{x}_1 = x_2, \quad \dot{x}_2 = -x_1 + \varepsilon \left(\frac{x_2^3}{3} - x_2 \right) \tag{3.50}$$

Let us choose as a Lyapunov function

$$V = \tfrac{1}{2}(x_1^2 + x_2^2) \tag{3.51}$$

which is clearly positive definite. Its derivative is

$$\dot{V} = x_1 \dot{x}_1 + x_2 \dot{x}_2$$

$$= x_1 x_2 - x_1 x_2 + \varepsilon \left(\frac{x_2^4}{3} - x_2^2 \right) \tag{3.52}$$

$$= -\varepsilon x_2^2 \left(1 - \frac{x_2^2}{3} \right)$$

Then if the parameter ε is positive, stability is asymptotic for $x_2^2 < 3$.

EXAMPLE 3.3.4: For the system given by

$$\dot{x}_1 = x_2 - a(x_1^2 + x_2^2)x_1$$

$$\dot{x}_2 = -x_1 - a(x_1^2 + x_2^2)x_2 \tag{3.53}$$

if we choose as a Lyapunov function

$$V_1 = \tfrac{1}{2}(x_1^2 + x_2^2) \tag{3.54}$$

it is easily found that

$$\dot{V}_1 = -a(x_1^2 + x_2^2)^2 \tag{3.55}$$

As an alternate approach, let

$$V_2 = \tfrac{1}{2}(\dot{x}_1^2 + \dot{x}_2^2) \tag{3.56}$$

By direct substitution from equations (3.53) we can show that

$$V_2 = V_1 + 4a^2 V_1^2 \tag{3.57}$$

which is clearly positive definite, as V_1 is itself definite (see equation (3.54)). Then the time derivative of V_2 is

$$\dot{V}_2 = (1 + 8a^2 V_1)\dot{V}_1 < 0 \tag{3.58}$$

so that both Lyapunov functions yield the identical result, i.e., that the origin is stable for $a > 0$, and not stable for $a < 0$, for the system (3.53). What makes this example really interesting is the fact that setting $a = 0$ in equations (3.53) yields a simple (linear) harmonic oscillator, whose equilibrium point is a center.

EXAMPLE 3.3.5: We will consider here a nonlinear system (c.g., a nonlinear spring with damping) with two equilibrium points. For positive values of the parameters a, b, we consider the stability of solutions of

$$\ddot{y} + a\dot{y} + 2by + 3y^2 = 0 \tag{3.59}$$

or, in first order form,

$$\dot{x}_1 = x_2, \quad \dot{x}_2 = -ax_2 - 2bx_1 - 3x_1^2 \tag{3.60}$$

The two equilibrium points are (see figure 5)

$$x_1 = 0, \quad x_2 = 0; \quad x_1 = \frac{-2b}{3}, \quad x_2 = 0 \tag{3.61}$$

First let us examine the linearized systems at the two equilibrium points. At $x_1 = x_2 = 0$, it is easy enough to find the characteristic equation in the form

$$\lambda^2 + a\lambda + 2b = 0 \tag{3.62}$$

which implies that the product and sum of the roots are positive and negative, respectively. Thus the origin must be either a stable node or a stable focus, but at any rate a point of asymptotic stability.

The linearized system at the other equilibrium point can be found by the transformation

$$y_1 = x_1 + \frac{2b}{3}, \quad y_2 = x_2 \tag{3.63}$$

which yields

$$\dot{y}_1 = y_2, \quad \dot{y}_2 = -ay_2 + 2by_1 \tag{3.64}$$

The characteristic roots of the system (3.64) are such that

$$\lambda_1 + \lambda_2 = -a, \quad \lambda_1\lambda_2 = -2b \tag{3.65}$$

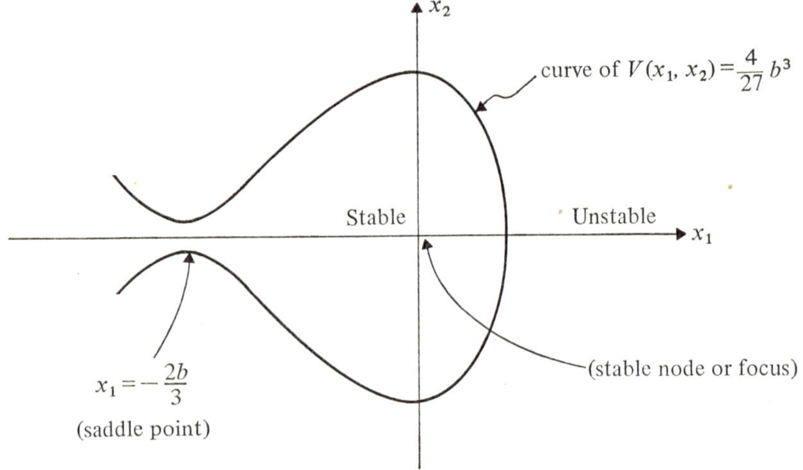

Fig. 5. *Phase plane plot for equations (3.59, 3.60), after LaSalle and Lefschetz [4].*

Thus the point $x_1 = -2b/3$, $x_2 = 0$ must be a saddle point, which is of course a condition of instability.

For both these equilibrium points, since neither is a center, a nonlinear analysis will yield no additional information about the local stability of equilibrium. However, we can use a Lyapunov function to try and get a stability boundary for the original system in the x_1, x_2 plane. That is, there is some region about $x_1 = x_2 = 0$, a region which is bounded by some stability boundary. Thus let us consider the function

$$V = \frac{1}{2}x_2^2 + x_1^2(x_1 + b) \tag{3.66}$$

which will be positive definite if $x_1 > -b$. Since the saddle point is at $x_1 = -2b/3$, this is satisfactory. The time rate of change of the Lyapunov function (3.66) can then be calculated as

$$\dot{V} = -ax_2^2 \tag{3.67}$$

which is clearly negative definite. Now at the point $x_2 = 0$, $x_1 = -2b/3$, we can evaluate V to be equal to $4b^3/27$, so that if we plotted on the x_1, x_2 plane the curve (figure 5)

$$V = bx_1^2 + x_1^3 + \frac{1}{2}x_2^2 = \frac{4}{27}b^3 \tag{3.68}$$

we could then state that any trajectory originating within this curve would tend asymptotically to the origin. Thus we have defined a stability region within the curve (3.68).

EXAMPLE 3.3.6: We now consider central-force motion in a Newtonian gravitational field. The appropriate equations are [2,8]:

$$m(\ddot{r} - r\dot{\phi}^2) = f(r) = -\frac{mK}{r^n}$$
$$r\ddot{\phi} + 2\dot{r}\dot{\phi} = 0 \tag{3.69}$$

The second of these equations can be integrated immediately to yield

$$r^2\dot{\phi} = h = \text{const.} \tag{3.70}$$

3 Autonomous system stability

We shall investigate in particular the stability of the motion described by $r = r_0 = \text{const.}$ From equation (3.70) we obtain for the ϕ variation in time the result

$$\phi = \frac{h}{r_0^2} t + c \tag{3.71}$$

where c is a constant of integration. The first of equations (3.69) then yields

$$- m r_0 \dot\phi^2 = f(r_0)$$

or

$$- \frac{h^2}{r_0^3} = \frac{1}{m} f(r_0) \tag{3.72}$$

which determines the constant h.

Now let us perturb the radius, that is let $r = r_0 + \xi(t)$, so that the differential equations of motion become

$$m(\ddot\xi - (r_0 + \xi)\dot\phi^2) = - \frac{mK}{(r_0 + \xi)^n}$$

$$\dot\phi = \frac{h}{(r_0 + \xi)^2} \tag{3.73}$$

We can eliminate $\dot\phi$ between these two equations to yield a single equation in ξ and its derivatives, i.e.,

$$m\ddot\xi - mh^2 (r_0 + \xi)^{-3} = f(r_0 + \xi)$$

$$= - \frac{mK}{(r_0 + \xi)^n} \tag{3.74}$$

To obtain the *linearized* system, we can expand the above differential equation in powers of ξ,

$$m\ddot\xi - \frac{mh^2}{r_0^3} \left(1 - \frac{3\xi}{r_0} + \frac{6\xi^2}{r_0^2} \cdots \right)$$

$$= f(r_0) + \xi f'(r_0) + \tfrac{1}{2}\xi^2 f''(r_0) + \cdots \tag{3.75}$$

If only the linear terms are retained above we obtain the equation

$$m\ddot{\xi} + \left(\frac{3mh^2}{r_0^4} - f'(r_0)\right)\xi = \frac{mh^2}{r_0^3} + f(r_0) \tag{3.76}$$

In view of the definition (3.72) of h, and the definition (3.74) of $f(r_0 + \xi)$, we can show that equation (3.76) can be reduced to

$$\ddot{\xi} + \left(\frac{3K}{r_0^{n+1}} - \frac{nK}{r_0^{n+1}}\right)\xi = 0$$

or

$$\ddot{\xi} + \frac{K}{r_0^{n+1}}(3 - n)\xi = 0 \tag{3.77}$$

Clearly for $n > 3$ the solutions to equation (3.77) are unstable, while for $n < 3$ we will get a center. For $n = 3$ the result is a repeated trivial root, which would lead to an instability. The unstable results ($n > 3$) would of course remain so for the nonlinear system, while the center cases ($n < 3$) will have to be investigated for the full nonlinear system.

If we define $x_1 = \xi$, and $x_2 = \dot{x}_1$, then the nonlinear system (3.74) can be written as

$$\dot{x}_1 = x_2, \quad \dot{x}_2 = h^2(r_0 + x_1)^{-3} - K(r_0 + x_1)^{-n} \tag{3.78}$$

As a Lyapunov function we shall choose

$$V = \frac{1}{2}\left(x_2^2 + \frac{h^2}{(r_0 + x_1)^2} - \frac{h^2}{r_0^2}\right) - \left(\frac{K}{1-n}\right)\left(r_0^{1-n} - (r_0 + x_1)^{1-n}\right) \tag{3.79}$$

It is easily seen that $V(0,0) = 0$, and it can be shown too that $\dot{V}(x_1, x_2) = 0$ for all (x_1, x_2). The remaining property required for V to be a Lyapunov function is that it be positive definite in some neighborhood of the origin. To find out where V will be definite, we will examine a Taylor series expansion of $V(x_1, 0)$, noting the symmetry of V with the variable x_2. Then

$$V(x_1, 0) = V(0, 0) + \left.\frac{\partial V(x_1, 0)}{\partial x_1}\right|_{x_1=0} x_1 + \frac{1}{2}\left.\frac{\partial^2 V(x_1, 0)}{\partial x_1^2}\right|_{x_1=0} x_1^2 + \ldots \tag{3.80}$$

We have already noted that $V(0, 0) = 0$, and then

$$\frac{\partial V(x_1, 0)}{\partial x_1}\bigg|_{x_1=0} = \frac{h^2}{(r_0 + x_1)^3} - \frac{K}{(r_0 + x_1)^n}\bigg|_{x_1=0} = 0 \qquad (3.81)$$

by virtue of the definition of h. Finally

$$\frac{\partial^2 V(x_1, 0)}{\partial x_1^2}\bigg|_{x_1=0} = \frac{3h^2}{(r_0 + x_1)^4} - \frac{nK}{(r_0 + x_1)^{n+1}}\bigg|_{x_1=0} = \frac{K}{r_0^{n+1}}(3-n) \qquad (3.82)$$

This number is positive only if $n < 3$, and so the (linearized) center result is shown to be stable according to the Lyapunov criterion.

We note in closing this section that we have not given any substantive guidelines on the construction of Lyapunov functions. The only systematic approach of any generality is a construction technique for linear autonomous systems, for which other stability criteria are also available. We have also noted that for mechanical systems there is a strong relation between Lyapunov functions and energy integrals, and this may be exploited in appropriate situations. The only other suggestion is to consult the literature for specific cases (see, for examples, Refs. [9-11]).

3.4   The theorem of minimum potential energy

As a final exercise in the discussion of Lyapunov functions for discrete systems (ordinary differential equations) we delve into Lagrangian and Hamiltonian mechanics to prove the famous theorem of *Lagrange*, to the effect that a system whose potential energy has a minimum is, in fact, stable. The theorem was first proved by Dirichlet, and later by Lyapunov, who used the Hamiltonian as a Lyapunov function. The present discussion is adapted from Meirovitch's extensive exposition [3].

We begin by recalling the Lagrangian and the Rayleigh dissipation function:

$$L = Lagrangian = T - V = Kinetic\ Energy\ -\ Potential\ Energy$$

$$F = Rayleigh\ dissipation\ function = \frac{1}{2}C_{rs}\dot{q}_r\dot{q}_s \qquad (3.83)$$

Then (obtained from a variational process) we can write down the well-known Lagrange Equations, which are the equivalent of (Newton's) equations of motion:

$$\frac{\mathrm{d}}{\mathrm{d}t}\left(\frac{\partial L}{\partial \dot{q}_k}\right) - \frac{\partial L}{\partial q_k} + \frac{\partial F}{\partial \dot{q}_k} = 0 \tag{3.84}$$

This produces a system of n (for a system with n degrees of freedom) second order differential equations in the generalized coordinates $q_i(t)$.

The Hamiltonian approach is an attempt to simplify the equations of motion, to *canonical equations*, by using what is really a Legendre transformation to reduce the equations to a system of first order equations.

We define *generalized momenta, p_i*:

$$p_i = \frac{\partial L}{\partial \dot{q}_i} \tag{3.85}$$

Further we define the *Hamiltonian H*:

$$H = \frac{\partial L}{\partial \dot{q}_k}\dot{q}_k - L = p_k\dot{q}_k - L = H(p_i, q_i) \tag{3.86}$$

Note that $H = H(p_i, q_i)$ is in contrast to $L = L(q_i, \dot{q}_i)$. From the definitions (3.85) and (3.86) it follows that

$$\delta H - p_k\delta\dot{q}_k + \dot{q}_k\delta p_k - \frac{\partial L}{\partial q_k}\delta q_k - \frac{\partial L}{\partial \dot{q}_k}\delta\dot{q}_k \tag{3.87}$$

Also, however,

$$\delta H = \frac{\partial H}{\partial p_i}\delta p_i + \frac{\partial H}{\partial q_i}\delta q_i \tag{3.88}$$

Thus by comparison of the variations (3.87) and (3.88) we see that

$$\dot{q}_k = \frac{\partial H}{\partial p_k} \tag{3.89}$$

From the Lagrange equations, however, in the absence of dissipative forces (the extension can be made), it can be seen that

$$\frac{\partial L}{\partial q_i} = \frac{\mathrm{d}}{\mathrm{d}t}\left(\frac{\partial L}{\partial \dot{q}_i}\right) = \frac{\mathrm{d}}{\mathrm{d}t}(p_i) = \dot{p}_i$$

or

$$\dot{p}_i = -\frac{\partial H}{\partial q_i} \tag{3.90}$$

The two equations (3.89) and (3.90) are the *Hamilton canonical equations.* For a system with a dissipation function,

$$\dot{p}_k = -\frac{\partial H}{\partial q_k} - \frac{\partial F}{\partial \dot{q}_k} \tag{3.91}$$

Note that the canonical equations are simply a set of first order ordinary differential equations.

Let us now compute $\mathrm{d}H/\mathrm{d}t$. By assumption, time does not appear explicitly in H (or even in L for that matter). Hence

$$\frac{\partial H}{\partial t} = -\frac{\partial L}{\partial t} = 0 \tag{3.92}$$

Then it follows that

$$\frac{\mathrm{d}H}{\mathrm{d}t} = \frac{\partial H}{\partial p_k}\dot{p}_k + \frac{\partial H}{\partial q_k}\dot{q}_k = 0 \tag{3.93}$$

where the last result must be true, from the canonical equations. Thus, if H were positive definite, with $\dot{H} = 0$, the Hamiltonian is a candidate to be a Lyapunov function for mechanical systems!

Now

$$H = p_i\dot{q}_i - L = p_i\dot{q}_i - (T - V) \tag{3.94}$$

Also, if the potential energy is independent of velocities,

$$p_i = \frac{\partial L}{\partial \dot{q}_i} = \frac{\partial T}{\partial \dot{q}_i}$$

so that

$$H = \frac{\partial T}{\partial \dot{q}_i} \dot{q}_i - T + V \tag{3.95}$$

If T is a homogeneous quadratic function of the velocities, then Euler's theorem can be applied, so that

$$\frac{\partial T}{\partial \dot{q}_i} \dot{q}_i = 2T \tag{3.96}$$

Hence

$$H = T + V = E = \text{total energy} \tag{3.97}$$

To arrive at this result, recall, we assumed a conservative system, with time-independent constraints, (see [12]), and with H not depending explicitly on time. Also, by definition, T, the kinetic energy, is positive definite.

Now we wish to examine the stability of the origin, $p_k = q_k = 0$, using the Hamiltonian system. If the potential energy is positive definite, then $H = T + V$ will be positive definite. Further, if V is positive definite, V has a minimum only at the origin. Thus, with $\dot{H} = 0$, H and V being positive definite, we can state Lagrange's Theorem:

THEOREM 3.7 (Lagrange): *If the potential energy has a (local) isolated minimum in the equilibrium position, the equilibrium is stable.*

We can also demonstrate the converse theorem (due to Lyapunov). Select a Lyapunov function

$$U = p_i q_i \tag{3.98}$$

Hence,

$$\dot{U} = \dot{p}_i q_i + p_i \dot{q}_i = p_i \frac{\partial H}{\partial p_i} - q_i \frac{\partial H}{\partial q_i} \tag{3.99}$$

If $H(p_k, q_k) = T(p_k) + V(q_k)$, then

$$\dot{U} = p_i \frac{\partial T}{\partial p_i} - q_i \frac{\partial V}{\partial q_i} \tag{3.100}$$

Clearly, however, by the arguments used above for $T = T(\dot{q}_i)$, it follows that

$$\dot{U} = 2T - q_i \frac{\partial V}{\partial q_i} \tag{3.101}$$

To get a maximum at the origin for V, assume it to be negative definite, and again invoke Euler's theorem on homogeneous functions, then

$$\dot{U} = 2T - \sum_{r=2}^{s} r\, V_r\,(q_k) \tag{3.102}$$

where the V_r are homogeneous functions of degree r. Thus $\dot{U}$ is positive definite, and U can be of either sign so that instability of the origin results if V has a maximum there.

THEOREM 3.8 (Lyapunov): *If, for a Hamiltonian system, the potential energy has a maximum in the equilibrium position, the equilibrium is unstable.*

Now consider the case where there is *damping*, represented in a Rayleigh dissipation function. Then the second Hamilton equation changes, since (see equation (3.91))

$$\frac{\partial H}{\partial q_i} = -\frac{\partial L}{\partial q_i} = -\frac{\mathrm{d}}{\mathrm{d}t}\left(\frac{\partial L}{\partial \dot{q}_i}\right) - \frac{\partial F}{\partial \dot{q}_i} = -\frac{\mathrm{d}}{\mathrm{d}t}(p_i) - \frac{\partial F}{\partial \dot{q}_i}$$

so that

$$\dot{p}_i = -\frac{\partial H}{\partial q_i} - \frac{\partial F}{\partial \dot{q}_i} \tag{3.103}$$

We also note that if T still presumed to be only a function of velocities, and V of the coordinates only, we could still identify H with the sum $T + V$, which could still be positive definite, but which would not be the 'total energy'. We can then compute $\dot{H}$:

$$\dot{H} = \frac{\partial H}{\partial p_i}\dot{p}_i + \frac{\partial H}{\partial q_i}\dot{q}_i = \frac{\partial H}{\partial p_k}\left(-\frac{\partial H}{\partial q_k} - \frac{\partial F}{\partial \dot{q}_k}\right) + \frac{\partial H}{\partial q_i}\left(\frac{\partial H}{\partial p_i}\right)$$

$$= -\frac{\partial H}{\partial p_k}\frac{\partial F}{\partial \dot{q}_k} = -\frac{\partial F}{\partial \dot{q}_k}\dot{q}_k \tag{3.104}$$

If $F = \frac{1}{2}C_{rs}\dot{q}_r\dot{q}_s$, by Euler's theorem, or straightforward algebra,

$$\dot{H} = -2F \tag{3.105}$$

We now give the following definitions: *complete damping* exists when $\dot{H}$ is a negative definite function of the generalized velocities $\dot{q}_k$; *pervasive damping* exists when $\dot{H}$ is only a negative semi-definite function of the generalized velocities.

Then for complete damping we will find asymptotic stability if H is positive definite. Further, if H is negative definite, even with complete damping, and $H < 0$, we would still get instability. Thus damping makes stability asymptotic, but doesn't effect instability. This confirms earlier results on the regular and inverted pendula.

Similarly if H is negative definite, with pervasive damping, we again would get instability.

The last results are based on a theorem due to Barbasin and Krasovskii, which is an extension of Lyapunov's theorems [3]:

THEOREM 3.9: *If there exists for the system $\dot{x} = X(x)$ a positive (negative) definite function $V(x)$ whose total time derivative $\dot{V}(x)$ is negative (positive) semi-definite along every trajectory of the system, and if the set of points S at which $\dot{V}(x)$ is zero contains no non-trivial positive half trajectory $x(t), t > t_0$, then the trivial solution $x = 0$ is asymptotically stable.*

It is important to note in closing this section that the principal result is Lagrange's theorem. This is so because it allows us, for discrete systems, to use a static energy criterion to evaluate the equilibrium points of dynamic systems.

3.5 Summary

In this chapter we have tried to summarize the principles used in the stability investigations of discrete systems. Thus we began with the Nyquist and Routh-Hurwitz criteria for linear systems, and then detailed and applied the theorems of Lyapunov's direct method for general autonomous systems. The direct method of Lyapunov is of interest because stability investigations could be conducted without obtaining solutions to the (nonlinear) governing differential equations. It is also of particular interest to note the motivation for the Lyapunov function in the first (energy) integral of equations of motion of mechanical systems.

The major point of this chapter, however, has been the application to the equations describing (discrete) mechanical systems, and the elicitation therefrom of the theorem of Lagrange and Dirichlet. The minimum energy theorem provides us with the philosophic and mathematical basis for our investigation of the stability of elastic systems.

References

[1] W. Hahn: *Stability of Motion.* Springer-Verlag, Berlin, 1967.

[2] H. Leipholz: *Stability Theory.* Academic Press, New York, 1970.

[3] L. Meirovitch: *Methods of Analytical Dynamics.* McGraw-Hill, New York, 1970

[4] L. LaSalle and S. Lefschetz: *Stability by Liapunov's Direct Method.* Academic Press, New York, 1961.

[5] A. M. Liapunov: *Stability of Motion.* Academic Press, New York, 1966.

[6] N. Minorsky: *Nonlinear Oscillations.* Van Nostrand, Princeton, 1962.

[7] W. Hahn: *Theory and Application of Liapunov's Direct Method.* Prentice-Hall, Englewood Cliffs, 1963.

[8] I. H. Shames: *Engineering Mechanics, Vol. II: Dynamics,* Prentice-Hall, Englewood Cliffs, 1966.

[9] O. Gurel and M. M. Salah: A Survey of Methods of Constructing Liapunov Functions. *International Business Machines Corporation* Technical Report No. 320-2943, New York, February 1967.

[10] A. K. Newman: New Lyapunov Function for Nonlinear Time-Varying Systems. *Journal of Basic Engineering*, Vol. 84 No. June 1968, p. 208-212.

[11] O. Gurel and L. Lapidus: A Guide to the Generation of Liapunov Functions. *Industrial and Engineering Chemistry*, Vol. 61, No. 3, March 1969, p. 30-41.

[12] H. Goldstein: *Classical Mechanics.* Addison-Wesley Publishing, Reading, Mass. 1959.

4

Stability concepts for continuous systems

We have pointed out a number of times that the principal result of the preceding chapter was the Lagrange (Dirichlet) theorem relating stable equilibrium to a minimum of the potential energy. This result is valid for all discrete mechanical systems. Unfortunately, no rigorous counterpart is available for continuous systems [1-6]. That is, although the potential energy criterion is the most often used criterion of structural stability, there is no rigorous proof available that the application of this criterion yields stability in the sense of Lyapunov.

Our intent in this chapter is to indicate by example the application of Lyapunov functionals for analyzing the stability of motion of continuous systems — in analogy to the Lyapunov functions for discrete systems — and we shall also record the principal stability theorems. Then we shall try to summarize the progress that has been made in linking the stability by Lyapunov functionals to the energy criterion, and, of course, we shall lay out various versions of the energy criterion for continuous elastic systems.

4.1 Lyapunov functionals for column stability

You will recall from our discussion of the stability of discrete systems that the concept of a Euclidian norm was used as a measure of length both in the definitions of stability and in the outline of the proofs of the Lyapunov stability theorems. Also, we used Lyapunov functions which were simply functions of the vector x with certain properties. To extend this approach (Lyapunov's direct method) we shall introduce the concepts of *metrics* and of *Lyapunov functionals* which are the counterparts of norms and Lyapunov functions for continuous systems. They will represent certain integrated or averaged values of the variables, over the spatial domain of the body or structure of interest. We shall state formal definitions and proofs in the next section. In this section we shall present some examples of column stability behavior, following the work of Plaut [7]. Although we have not derived the equations for the dynamic (or static) behavior of columns — as we will in Chapters 5 and 6 — we shall assume that these equations are known.

The governing partial differential equation for the transverse deflection W of a column loaded by a compressive axial force P is

$$EI \frac{\partial^4 W}{\partial X^4} + P \frac{\partial^2 W}{\partial X^2} + \rho A \frac{\partial^2 W}{\partial T^2} = 0 \tag{4.1}$$

where EI is the bending stiffness, ρ the mass density, A the area, and X and T are the axial and time coordinates, respectively. Note that $0 \leq X \leq L$, and $T > 0$. If we introduce the dimensionless parameters

$$w = \frac{W}{L}, \quad x = \frac{X}{L}, \quad t = T \sqrt{\frac{EI}{\rho A L^4}}$$

$$p = \frac{PL^2}{EI} = \frac{\pi^2 P}{P_E}, \quad P_E = \frac{\pi^2 EI}{L^2} \tag{4.2}$$

where P_E is the classical Euler load of a simply supported column, then the governing equation can be written as

$$\frac{\partial^2 w}{\partial t^2} + \frac{\partial^4 w}{\partial x^4} + p \frac{\partial^2 w}{\partial x^2} = 0 \tag{4.3}$$

Then if the velocity $v = v(x, t)$ is introduced, we can write equation (4.3) as a system of equations of first order in the time derivative, i.e.,

$$\frac{\partial w}{\partial t} = v$$

$$\frac{\partial v}{\partial t} = -\frac{\partial^4 w}{\partial x^4} - p \frac{\partial^2 w}{\partial x^2} \tag{4.4}$$

It is then also convenient to introduce the notation of a vector $\boldsymbol{u}$ defined by

$$u = \begin{pmatrix} w \\ v \end{pmatrix} \tag{4.5}$$

Then we can say that the vector $u = 0$ corresponds to the straight line equilibrium position.

Let us now introduce the *functional*

$$V_1(u) = \int_0^1 \left[v^2 + \left(\frac{\partial^2 w}{\partial x^2} \right)^2 - p \left(\frac{\partial w}{\partial x} \right)^2 \right] dx \tag{4.6}$$

which actually represents twice the total kinetic and potential energy of the deformed column (above the equilibrium state $u = 0$). A functional, mathematically speaking, is a function of a function. Thus the dependence of $V_1(u)$ upon u, that is the number that results from the integration in equation (4.6), depends upon the particular vector that is used to calculate the integrand. In analogy to our work on Lyapunov functions, we shall investigate below the sign of $V_1(u)$ and of its time derivative.

Let us also introduce the auxiliary vectors ξ and η as

$$\xi = \begin{pmatrix} \xi_1 \\ \xi_2 \end{pmatrix}, \quad \eta = \begin{pmatrix} \eta_1 \\ \eta_2 \end{pmatrix} \tag{4.7}$$

so that we can introduce also the *metric* $\rho_1(\xi, \eta)$,

$$\rho_1(\xi, \eta) = \left\{ \int_0^1 \left[(\xi_2 - \eta_2)^2 + \left(\frac{\partial^2 (\xi_1 - \eta_1)}{\partial x^2} \right)^2 \right. \right.$$
$$\left. \left. + \left(\frac{\partial (\xi_1 - \eta_1)}{\partial x} \right)^2 + (\xi_1 - \eta_1)^2 \right] dx \right\}^{1/2} \tag{4.8}$$

It then follows that

$$\rho_1(u, 0) = \left\{ \int_0^1 \left[v^2 + \left(\frac{\partial^2 w}{\partial x^2} \right)^2 + \left(\frac{\partial w}{\partial x} \right)^2 + w^2 \right] dx \right\}^{1/2} \tag{4.9}$$

Thus $\rho_1(u, 0)$ is a measure of the 'distance' between the equilibrium state $u = 0$ and the deformed state $u = u$. Further, if $\rho_1(u, 0)$ is small, then each of the terms

$$\int_0^1 v^2 \, dx, \quad \int_0^1 \left(\frac{\partial^2 w}{\partial x^2}\right)^2 dx, \quad \int_0^1 \left(\frac{\partial w}{\partial x}\right)^2 dx, \quad \int_0^2 w^2 \, dx$$

must then also be small, as the integrand of $\rho_1(\boldsymbol{u}, \boldsymbol{0})$ is the sum of these non-negative terms.

Note that the metric $\rho_1(\boldsymbol{\xi}, \boldsymbol{\eta})$ has the following properties,

$$\rho_1(\boldsymbol{\xi}, \boldsymbol{\eta}) \geq 0, \quad \rho_1(\boldsymbol{\xi}, \boldsymbol{\xi}) = 0$$

$$\rho_1(\boldsymbol{\xi}, \boldsymbol{\eta}) = \rho_1(\boldsymbol{\eta}, \boldsymbol{\xi}) \tag{4.10}$$

and also

$$\rho_1(\boldsymbol{\xi}, \boldsymbol{\eta}) \leq \rho_1(\boldsymbol{\xi}, \boldsymbol{\phi}) + \rho_1(\boldsymbol{\phi}, \boldsymbol{\eta}) \tag{4.11}$$

where $\boldsymbol{\phi}$ is also a two-dimensional vector.

We had noted in Chapter 3 that the choice of a Lyapunov function is not unique. It is true here not only for the choice of Lyapunov functional, but also for the choice of metric. Thus it would also be possible to use metrics such as

$$\rho_2(\boldsymbol{u}, \boldsymbol{0}) = \left\{ \int_0^1 \left[v^2 + \left(\frac{\partial^2 w}{\partial x^2}\right)^2 + \left(\frac{\partial w}{\partial x}\right)^2 \right] dx \right\}^{1/2} \tag{4.12}$$

$$\rho_3(\boldsymbol{u}, \boldsymbol{0}) = \left\{ \int_0^1 \left[v^2 + \left(\frac{\partial^2 w}{\partial x^2}\right)^2 \right] dx \right\}^{1/2} \tag{4.13}$$

$$\rho_4(\boldsymbol{u}, \boldsymbol{0}) = \sup_x | w | + \left\{ \int_0^1 \left[v^2 + \left(\frac{\partial^2 w}{\partial x^2}\right)^2 \right] dx \right\}^{1/2} \tag{4.14}$$

where $\sup_x | w |$ is the least upper bound of $| w(x, t) |$ for $0 \leq x \leq 1$. We shall state in the next section the formal conditions for the *equivalence of metrics*.

To prove the stability of the equilibrium state $\boldsymbol{u} = \boldsymbol{0}$, in analogy with our work on discrete systems, we proceed by finding the conditions under which $V_1(\boldsymbol{u})$ is positive definite, and then we investigate the sign of $\dot{V}_1(\boldsymbol{u})$. As $\rho_1^2(\boldsymbol{u}, \boldsymbol{0})$ is clearly positive definite, we can prove the positivity of $V_1(\boldsymbol{u})$ if we can show that

$$V_1(\boldsymbol{u}) \geq \alpha \rho_1^2(\boldsymbol{u}, \boldsymbol{0}) \tag{4.15}$$

where α is a positive constant.

To accomplish this task it is useful to note the following inequalities [7-10]:

$$\text{If } w(0) = w(1) = 0, \quad \int_0^1 \left(\frac{\partial w}{\partial x}\right)^2 dx \geq \pi^2 \int_0^1 w^2 \, dx \tag{4.16}$$

$$\text{If } w(0) = w(1), \text{ and if } \frac{\partial w(0)}{\partial x} = \frac{\partial w(1)}{\partial x} = 0,$$

$$\int_0^1 \left(\frac{\partial^2 w}{\partial x^2}\right)^2 dx \geq 4\pi^2 \int_0^1 \left(\frac{\partial w}{\partial x}\right)^2 dx \tag{4.17}$$

$$\text{If } w(0) = w(1), \text{ and if } \frac{\partial^2 w(0)}{\partial x^2} = \frac{\partial^2 w(1)}{\partial x^2} = 0,$$

$$\int_0^1 \left(\frac{\partial^2 w}{\partial x^2}\right)^2 dx \geq \pi^2 \int_0^1 \left(\frac{\partial w}{\partial x}\right)^2 dx \tag{4.18}$$

These and other inequalities can be obtained through variational statements of some elementary eigenvalue problems. It is interesting to note the dependence of these inequalities on the boundary conditions that must be satisfied by $w(x, t)$.

To prove the positivity of $V_1(\boldsymbol{u})$, we rewrite equation (4.6) in the form

$$V_1(\boldsymbol{u}) = \int_0^1 \left[v^2 + (1 - c)\left(\frac{\partial^2 w}{\partial x^2}\right)^2 + c\left(\frac{\partial^2 w}{\partial x^2}\right)^2 - p\left(\frac{\partial w}{\partial x}\right)^2 \right] dx \tag{4.19}$$

where $0 < c < 1$. If we considered columns that are pinned, we could then use inequality (4.18) to note that

$$V_1(\boldsymbol{u}) \geq \int_0^1 \left[v^2 + (1 - c)\left(\frac{\partial^2 w}{\partial x^2}\right)^2 + (c\pi^2 - p)\left(\frac{\partial w}{\partial x}\right)^2 \right] dx \tag{4.20}$$

We shall assume that $p < \pi^2$. Then we choose c such that

$$c = \frac{p + \pi^2}{2\pi^2} \tag{4.21}$$

from which it follows that

$$1 - c = \frac{\pi^2 - p}{2\pi^2} > 0, \quad c\pi^2 - p = \frac{\pi^2 - p}{2} > 0 \tag{4.22}$$

Thus the inequality (4.20) can be written as

$$V_1(\boldsymbol{u}) \geq \int_0^1 \left[v^2 + \left(\frac{\pi^2 - p}{2\pi^2} \right) \left(\frac{\partial^2 w}{\partial x^2} \right)^2 + \left(\frac{\pi^2 - p}{4} \right) \left(\frac{\partial w}{\partial x} \right)^2 \right.$$
$$\left. + \left(\frac{\pi^2 - p}{4} \right) \left(\frac{\partial w}{\partial x} \right)^2 \right] \mathrm{d}x \tag{4.23}$$

which can be further manipulated with the aid of the inequality (4.16) to yield

$$V_1(\boldsymbol{u}) \geq \int_0^1 \left[v^2 + \left(\frac{\pi^2 - p}{2\pi^2} \right) \left(\frac{\partial^2 w}{\partial x^2} \right)^2 + \left(\frac{\pi^2 - p}{4} \right) \left(\frac{\partial w}{\partial x} \right)^2 + \right.$$
$$\left. + \frac{(\pi^2 - p)\pi^2}{4} w^2 \right] \mathrm{d}x \tag{4.24}$$

Clearly, then, if we choose $\alpha = (\pi^2 - p)/2\pi^2$, we can satisfy the condition (4.15) and verify that $V_1(\boldsymbol{u})$ is positive definite.

It is also necessary, for the Lyapunov functionals, to show that $V_1(\boldsymbol{u})$ admits an infinitely small upper bound [1] in the neighborhood of the equilibrium state $\boldsymbol{u} = \boldsymbol{0}$. Equivalently we could require that

$$V_1(\boldsymbol{u}) \leq \gamma\rho_1^2(\boldsymbol{u}, \boldsymbol{0}) \tag{4.25}$$

where γ is a positive constant. By a simple comparison of equations (4.6) and (4.9) we can see that if $\gamma = \max\{1, p\}$ we can satisfy the inequality (4.25).

---

[1] The reason for this is to allow continuity of the solution of the partial differential equation (4.4) with respect to the initial data. Thus by limiting the size of the initial disturbance, a bound can be placed on the response. Some authors use a separate metric, ρ_0 say, for this purpose, and then require that $\rho_1^2 < \text{const.} \times \rho_0^2$.

Finally we need to examine the time derivative of $V_1(\mathbf{u})$. By straightforward differentiation

$$\frac{\mathrm{d}V_1}{\mathrm{d}t} = 2\int_0^1 \left[v\,\frac{\partial v}{\partial t} + \frac{\partial^2 w}{\partial x^2}\,\frac{\partial^2 v}{\partial x^2} - p\,\frac{\partial w}{\partial x}\,\frac{\partial v}{\partial x} \right]\mathrm{d}x \qquad (4.26)$$

wherein we have used the first of equations (4.4). If we integrate the above integrand by parts, integrating the spatial derivatives of the velocity, we find that

$$\frac{\mathrm{d}V_1}{\mathrm{d}t} = 2\int_0^1 \left[\frac{\partial v}{\partial t} + \frac{\partial^4 w}{\partial x^4} + p\,\frac{\partial^2 w}{\partial x^2} \right] v\,\mathrm{d}x$$

$$+\ 2\left[\frac{\partial^2 w}{\partial x^2}\,\frac{\partial v}{\partial x} - \frac{\partial^3 w}{\partial x^2}v - p\,\frac{\partial w}{\partial x}v \right]_{x=0}^{x=1} \qquad (4.27)$$

As the motion must satisfy both the differential equation of motion, and the boundary conditions, it is seen that $\dot{V}_1(\mathbf{u}) = 0$. Thus, by a theorem yet to be stated, with the assumption

$$p < \pi^2 \quad \text{or} \quad P < P_E \qquad (4.28)$$

we can state that the column would be stable for loads satisfying the inequality (4.28).

It is also possible [1] (see Plaut [7]) to show that for $p > \pi^2$, using Lyapunov type instability theorems for Lyapunov functionals, the equilibrium state $\mathbf{u} = 0$ is not stable.

We shall not pursue this approach any further here.[2] In addition to the thorough work of Plaut already cited, applications have been presented by Movchan [11, 12], Parks [13], Wang [14, 15], Hegemier [16], and Berger and Lapidus [17]. The theory for continuous systems is largely due to Zubov [18] and Movchan [11, 12]. We shall record the principal results of that theory in the following section.

We might also note here that the direct method of Lyapunov has also

---

been effectively used to calculate displacement and energy bounds for continuous systems by Plaut [19] and Holzer [20, 21].

4.2 Definitions and theorems for Lyapunov functionals

We will consider[1] a set of $2n$ partial differential equations of the form

$$\frac{\partial u_i}{\partial t} = g_i\left(x_j,\ u_m,\ \frac{\partial u_m}{\partial x_j},\ \ldots,\ \frac{\partial^{(\gamma_i)} u_m}{\partial x_j^{(\gamma_i)}}\right),\quad \begin{array}{l}(i = 1, 2, \ldots, 2n)\\(m = 1, 2, \ldots, 2n)\end{array} \tag{4.29}$$

where $x_1, \ldots, x_k$ are spatial coordinates and t is the time variable. The highest order of derivatives appearing in each equation is denoted by γ_i. We take the equations to be autonomous, i.e., the functions g_i do not depend explicitly on t. The variables $u_1, \ldots, u_n$ are generally associated with displacement components, while $u_{n+1}, \ldots, u_{2n}$ will be associated with velocity components.

Equations (4.29) will apply in a certain region D of the k-dimensional Euclidian space E_k. For $k = 1$ there is a single spatial coordinate x_1 and we have a 'one-dimensional' system. For $k = 2$, then the equations are applicable in a region of the $x_1 x_2$ plane and we have a 'two-dimensional' system.

In addition to equations (4.29), a set of appropriate boundary conditions and smoothness conditions will be specified. These conditions will be taken to be independent of time (sometimes called stationary conditions), so that the systems will be autonomous. The smoothness conditions refer to differentiability and continuity conditions for $u_1, \ldots, u_{2n}$ in D, of the form

$$u_i \in C^{p_i},\quad (i = 1, 2, \ldots, 2n) \tag{4.30}$$

where each p_i is a positive integer and C^{p_i} is the set of real functions $f(x_1, \ldots, x_k)$ whose partial derivatives of order $\leq p_i$ all exist and are continuous in D. Equations (4.29) together with the conditions above will be referred to as the *boundary value problem*.

Consider the *functional space* Φ composed of real vector elements

$$\phi = \begin{pmatrix} \phi_1 \\ \vdots \\ \phi_{2n} \end{pmatrix} \tag{4.31}$$

---

[1] Our exposition here follows Plaut [7].

whose components $\phi_1, \ldots, \phi_{2n}$ satisfy the boundary conditions and the smoothness conditions. $\phi_1, \ldots, \phi_{2n}$ are functions of the spatial coordinates. We can think of elements $\phi \in \Phi$ as *admissible states* for the system. Let

$$u = \begin{pmatrix} u_1 \\ \vdots \\ u_{2n} \end{pmatrix} \tag{4.32}$$

where the components u_i are the dependent variables of equations (4.29). We assume that for any element $u^0 \in \Phi$ there exists a unique solution $u = u(t, u^0)$ of the boundary value problem having the following properties:

i) $u(t, u^0)$ is defined for all $t \geq 0$; $\tag{4.33a}$

ii) $u(t, u^0)$ is continuous in t and u^0; $\tag{4.33b}$

iii) $u(0, u^0) = u^0$ $\tag{4.33c}$

In addition, we assume that

iv)   if $u^0 = 0$, then $u(t, 0) = 0$ for all $t \geq 0$ $\tag{4.33d}$

Since $u(t, u^0)$ is a solution of the boundary value problem,

v) $u(t, u^0) \in \Phi$ for all $t \geq 0$ $\tag{4.33e}$

and the components $u_i(t, u^0)$ satisfy equations (4.19). Property (iii) implies that u^0 represents the *initial state* of the system. We write

$$u^0 = \begin{pmatrix} u_1^0 \\ \vdots \\ u_{2n}^0 \end{pmatrix} \tag{4.34}$$

with $u_i^0 = u_i^0(x_1, \ldots, x_k)$. For fixed u^0, $u(t, u^0)$ is called the *motion* of the system. The motion satisfies the boundary conditions and smoothness conditions at each instant of time. The condition (4.33d) implies that the system has a trivial solution $u = 0$ $(u_1 = u_2 = \ldots = u_{2n} = 0)$. The equations of motion are formulated so that the solution $u = 0$ corresponds to the equilibrium configuration to be investigated. This allows us to put our discussion in terms of the stability of $u = 0$.

Now we introduce the following definitions:

1)   We say that a *metric* ρ is defined on Φ if to any two elements $\phi \in \Phi$ and $\psi \in \Phi$ there corresponds a real number $\rho(\phi, \psi)$ satisfying the following conditions:

a) $\rho(\phi, \psi) \geq 0;$ $\qquad\qquad\qquad\qquad\qquad\qquad\qquad\qquad$ (4.35a)

b) $\rho(\phi, \psi) = 0$ if and only if $\phi = \psi;$ $\qquad\qquad\qquad\qquad\qquad$ (4.35b)

c) $\rho(\phi, \psi) = \rho(\psi, \phi);$ $\qquad\qquad\qquad\qquad\qquad\qquad\qquad$ (4.35c)

d) $\rho(\phi, \psi) \leq \rho(\phi, \eta) + \rho(\eta, \psi)$ for any $\eta \in \Phi.$ $\qquad\qquad\quad$ (4.35d)

The space Φ is then called a *metric space*.

2)   Two metrics ρ_1 and ρ_2 are said to be *equivalent* if there exist constants $a > 0, b > 0$ such that

$$a\rho_1(u, 0) < \rho_2(u, 0) < b\rho_1(u, 0) \qquad\qquad\qquad (4.36)$$

for any $u \in \Phi.$

3)   Let d be a positive real number. The *neighborhood $S(0, d)$ of $u = 0$* is defined as the set of $u \in \Phi$ for which

$$0 \leq \rho(u, 0) < d. \qquad\qquad\qquad\qquad\qquad (4.37)$$

4)   The element $u^{(e)} \in \Phi$ is said to be an *equilibrium state* of the boundary value problem if

$$u(t, u^{(e)}) = u^{(e)} \quad \text{for all } t \geq 0 \qquad\qquad\qquad (4.38)$$

The equilibrium states are determined by setting the functions g_i in equations (4.29) equal to zero and solving the resulting set of equations. In studying the stability of $u = 0$, it will be assumed that $u = 0$ is an isolated equilibrium state, i.e., there exists a $\delta > 0$ such that $S(0, \delta)$ contains no equilibrium state other than $u = 0.$

Let ε and δ be real numbers. Referred to the metric space Φ, the following definitions of stability and instability will be considered:

5)   The solution $u = 0$ of the boundary value problem is said to be *stable with respect to ρ* if for every $\varepsilon > 0$ there exists a $\delta > 0$ such that

$$\rho(u^0, 0) < \delta \text{ implies that } \rho(u(t, u^0), 0) < \varepsilon \text{ for all } t \geq 0. \qquad (4.39)$$

If, in addition,

$$\rho(\boldsymbol{u}(t, \boldsymbol{u}^0), \boldsymbol{0}) \to 0 \text{ as } t \to +\infty, \tag{4.40}$$

then $\boldsymbol{u} = \boldsymbol{0}$ is said to be *asymptotically stable with respect to ρ*.

6) The solution $\boldsymbol{u} = \boldsymbol{0}$ of the boundary value problem is said to be *unstable with respect to ρ* if there exists an $\varepsilon > 0$ such that for any $\delta > 0$, no matter how small, it is possible to find a $\boldsymbol{u}^0 \in \Phi$ for which

$$\rho(\boldsymbol{u}^0, \boldsymbol{0}) < \delta \text{ and } \rho(\boldsymbol{u}(t, \boldsymbol{u}^0), \boldsymbol{0}) \geq \varepsilon \text{ for some } t > 0. \tag{4.41}$$

Note how these definitions parallel those of the discrete systems (Chapter 2).

7) We say the V is a *functional* defined on Φ if to each element $\boldsymbol{u} \in \Phi$ there corresponds a real number $V(\boldsymbol{u})$.

8) The functional V is said to be *positive definite with respect to ρ* if for any $\varepsilon > 0$ there exists a $\delta > 0$ such that

$$V(\boldsymbol{u}) > \delta \quad \text{whenever} \quad \rho(\boldsymbol{u}, \boldsymbol{0}) > \varepsilon \tag{4.42}$$

Note that V is positive definite with respect to ρ if

$$V(\boldsymbol{u}) \geq \alpha\rho^2(\boldsymbol{u}, \boldsymbol{0}), \quad \text{where } \alpha \text{ is a positive constant.} \tag{4.43}$$

9) The functional V is said to admit an *infinitely small upper bound with respect to ρ* if for any $\varepsilon > 0$, as small as desired, there exists a $\delta > 0$ such that

$$|V(\boldsymbol{u})| \leq \varepsilon \quad \text{whenever} \quad \rho(\boldsymbol{u}, \boldsymbol{0}) < \delta \tag{4.44}$$

Note that V admits an infinitely small upper bound with respect to ρ if

$$|V(\boldsymbol{u})| \leq \gamma\rho^2(\boldsymbol{u}, \boldsymbol{0}), \text{ where } \gamma \text{ is a positive constant} \tag{4.45}$$

10) The functional V is said to be *bounded* if there exists a positive constant M such that

$$|V(\boldsymbol{u})| < M \tag{4.46}$$

Thus again we see the parallel between the definitions of $V(\boldsymbol{x})$ for discrete systems given in Chapter 3, and the definitions of $V(\boldsymbol{u})$ above.

The principal theorems are then given as follows:

THEOREM 4.1 (Zubov): *In order for the solution $\mathbf{u} = \mathbf{0}$ of the boundary value problem to be stable with respect to ρ, it is necessary and sufficient that in a sufficiently small neighborhood $S(\mathbf{0}, d)$ of $\mathbf{u} = \mathbf{0}$ there exists a functional V having the following properties when $\mathbf{u} \in S(\mathbf{0}, d)$:*
1) *V is positive definite with respect to ρ;*
2) *V admits an infinitely small upper bound with respect to ρ;*
3) *$V(\mathbf{u}(t, \mathbf{u}^0))$ is non-increasing for $t \geq 0$, whenever $\mathbf{u}^0 \in S(\mathbf{0}, d)$.*
If, in addition, there exists a d', $0 < d' \leq d$, such that
4) *$V(\mathbf{u}(t, \mathbf{u}^0)) \to 0$ as $t \to +\infty$, whenever $\mathbf{u}^0 \in S(\mathbf{0}, d')$, then $\mathbf{u} = \mathbf{0}$ is asymptotically stable with respect to ρ.*

In practice we can show satisfaction of the conditions of the theorem in three steps, i.e.,
1') $V(\mathbf{u}) \geq \alpha\rho^2(\mathbf{u}, \mathbf{0})$, where α is a positive constant;
2') $| V(\mathbf{u}) | \leq \gamma\rho^2(\mathbf{u}, \mathbf{0})$, where γ is a positive constant;
3') $\dfrac{\mathrm{d}V(\mathbf{u}(t, \mathbf{u}^0))}{\mathrm{d}t} \leq 0$ for $t \geq 0$.

For conservative systems, we may choose V as the energy in the system, and then, by the principle of conservation of energy,

$$\frac{\mathrm{d}V(\mathbf{u}(t, \mathbf{u}^0))}{\mathrm{d}t} = 0 \tag{4.47}$$

If the equations of motion are linear, and if a metric ρ is chosen such that $\rho^2(\mathbf{u}, \mathbf{0})$ contains terms similar to those in V, then step 2' is satisfied and it remains to find the conditions under which step 1' holds. If the equations of motion are nonlinear, steps 1' and 2' are applied only to a neighborhood $S(\mathbf{0}, d)$ of $\mathbf{u} = \mathbf{0}$. For nonconservative systems we may not always be able to find an appropriate stability functional V.

THEOREM 4.2 (Movchan): *In order for the solution $\mathbf{u} = \mathbf{0}$ of the boundary value problem to be unstable with respect to ρ, it is necessary and sufficient that in a sufficiently small neighborhood $S(\mathbf{0}, d)$ of $\mathbf{u} = \mathbf{0}$ there exists a functional W having the following properties:*
1) *in as small a neighborhood of $\mathbf{u} = \mathbf{0}$ as desired, there exists at least one element $\mathbf{u}^0 \in \Phi$, $\mathbf{u}^0 \neq \mathbf{0}$, such that $W(\mathbf{u}^0) > 0$;*
2) *$W(\mathbf{u})$ is bounded, whenever $W(\mathbf{u}) > 0$ and $\mathbf{u} \in S(\mathbf{0}, d)$;*

3) $W(\boldsymbol{u}(t, \boldsymbol{u}^0))$ *is increasing for* $t \geq 0$, *whenever* $W(\boldsymbol{u}(t, \boldsymbol{u}^0)) > 0$ *and* $\boldsymbol{u}(t, \boldsymbol{u}^0) \in S(\boldsymbol{0}, d)$.

We can exhibit the following properties as the equivalent of showing instability:

1')   given $\delta > 0$, $\boldsymbol{u}^0$ is found for which $0 < \rho(\boldsymbol{u}^0, \boldsymbol{0}) < \delta$ and $W(\boldsymbol{u}^0) > 0$;

2') $|\,W(\boldsymbol{u})\,| \leq \gamma\rho^4(\boldsymbol{u}, \boldsymbol{0})$, where γ is a positive constant;

3') $\dfrac{\mathrm{d}W(\boldsymbol{u}(t, \boldsymbol{u}^0))}{\mathrm{d}t} > 0$ whenever $W(\boldsymbol{u}(t, \boldsymbol{u}^0)) > 0$.

Now we indicate that stability with respect to a metric ρ implies stability with respect to any metric which is equivalent to ρ, plus a similar theorem regarding instability. Proofs are given by Plaut [7].

THEOREM 4.3: *If the solution* $\boldsymbol{u} = \boldsymbol{0}$ *of the boundary value problem is stable with respect to* ρ, *then it is also stable with respect to any metric equivalent to* ρ.

THEOREM 4.4: *If the solution* $\boldsymbol{u} = \boldsymbol{0}$ *of the boundary value problem is unstable with respect to* ρ, *then it is also unstable with respect to any metric equivalent to* ρ.

4.3   The minimum potential energy criterion

For all the applications that we wish to consider, in the field of elastic stability, we will adopt the following view, as expressed by Hutchinson and Koiter [6]: 'The energy criterion of stability for elastic systems subject to conservative loading is almost universally accepted by workers in the field of structural stability. A positive definite second variation of the potential energy about a static equilibrium state is accepted as a sufficient condition for the stability of that state. Numerous attempts to undermine these two pillars of structural stability theory have been made, but confidence in them remains undiminished. With a proper shoring-up of certain aspects of these criteria, they will undoubtedly continue to serve as the foundation of elastic stability theory'.

The principal problem that is involved here has a number of parts, i.e., 1) how can the definition of Lyapunov stability for discrete systems be suitably extended for continuous systems; 2) with an appropriate stability definition for continuous systems, is it possible to lay down an equivalent for such systems to the Lagrange theorem for discrete systems; and, 3) can the criterion of a proper minimum of the potential energy for stability be translated, for linear and nonlinear systems, into the equivalent of Lyapunov stability.

These questions are important simply because, as implied by the quotation above, the criterion of Trefftz and the potential energy criterion have been the basis of almost all the structural stability investigations that are known. In fact these criteria have found widespread application because, in general, they lead to results that have been verified by experience. The principle exceptions to the correlation of theory and practice have been in the analysis of what are now termed *imperfection-sensitive structures*. Such structures are not stable at the critical or buckling load, (the equilibrium point), and their proper analysis requires the inclusion of nonlinear effects.

A complete and proper discussion of the intricacies involved in developing the energy criterion and the Trefftz criterion for continuous systems, within the framework of Lyapunov stability, is beyond our scope. The reader is referred to the works of Koiter [3-5] for a rigorous discussion. We wish only to outline briefly some of the problems encountered, and to state the principal results.

Part of the problem, as indicated, is in the extension of the definition of stability. It has been found (see, for example, Refs. [10, 5, 9]) that if the local displacements and velocities — on a point by point basis — are required to be as small as we please for all time, which is one analogue of requiring $||\, x\,|| < \varepsilon$ for discrete systems, then in some cases these requirements cannot be met no matter how small the initial disturbance. In some situations the natural state of zero stress and strain is unstable [4].

Koiter [4] has suggested requiring that the following average quantities (instead of the local displacements and velocities)

$$|| \, u \, ||^2 = \frac{1}{V} \int (u \cdot u) \, \mathrm{d}V, \quad || \, \dot{u} \, ||^2 = \frac{1}{V} \int (\dot{u} \cdot \dot{u}) \, \mathrm{d}V \tag{4.48}$$

be arbitrarily small. These norms (squared) of the displacement vector u and the velocity vector $\dot{u}$, averaged over the volumes of the bodies under consideration, are termed L_2-*norms*. Then an equilibrium configuration is considered stable if these two L_2-norms remain as small as we please for all (positive) time, and for sufficiently small initial disturbances that the initial energy imparted to the system is small. With this modified stability definition, which reduces quite naturally to the Lyapunov definition, Koiter has proved the following theorem [4, 5]:

THEOREM 4.5 (Koiter): *A proper minimum of the potential energy in the equilibrium configuration, in comparison with the energy in all other possible configurations characterized by a sufficiently small L_2-norm, is a sufficient condition for stability.*

The sufficient condition for stability can be written as

$$V(\boldsymbol{u}) \geq \mathrm{d}(c) > 0 \quad \text{for} \quad \| \boldsymbol{u} \| = c \tag{4.49}$$

where $V(\boldsymbol{u})$ is the potential energy, and c, d are small positive numbers. A *weak minimum* of the energy is

$$V(\boldsymbol{u}) \geq 0 \quad \text{for} \quad | u_i | < g, \quad | u_{i,j} | < h \tag{4.50}$$

where the u_i are components of $\boldsymbol{u}$, $u_{i,j}$ are the displacement gradients (strains), and g, h are small positive constants. Now a sufficient condition for a weak minimum is a positive-definite second variation of the energy,

$$\delta^{(2)} V(\boldsymbol{u}) > 0 \tag{4.51}$$

Note that the definition (4.50), and by extension the condition (4.51), depends on the u_i being small at all points, etc.; the condition (4.49) depends on the L_2-norm.

It was found, however, that without suitable modifications to the energy functional, the inequalities (4.50) and (4.51) do not ensure the stability sufficiency condition (4.49). The reason for this is that the strain gradients may be unbounded, and the classical theory of finite deformations disregards the possible effects of strain gradients on the strain energy. That is, the elastic energy depends only on the strains, and not on their gradients.

However, by consideration of a modified strain energy that includes strain gradient terms, and by also examining the effects of the damping which is always present in physical systems, Koiter was able to state the following result [5]:

THEOREM: 4.6 (Koiter): *The existence of a (weak) proper minimum of the potential energy in the equilibrium state constitutes, for all practical purposes, both a necessary and sufficient condition for the stability of this configuration in the dynamical sense of Lyapunov.*

Thus, for 'all practical purposes', we may now take the point of view that a minimum of the potential energy or (equivalently) a positive definite second variation of the energy implies stability in the sense of Lyapunov. This is the basis for all our work in the sequel on the stability of (static) conservative systems, although we shall have occasion to mention (Chapter 6) other criteria which have been used to examine structural stability.

We might also mention here an argument presented by Ziegler [1]. He points out — as shall become evident in the sequel — that whenever we actually set out to investigate the stability of an elastic structure, we generally apply some approximate technique. This technique, e.g., Rayleigh-Ritz or Galerkin, effectively discretizes our continuous system anyway. Thus, argues Ziegler, we could probably be content to simply apply the Lagrange theorem to the discretized versions of our continuous systems.

4.4 Summary

We have presented in this chapter a discursive approach to Lyapunov stability of continuous systems. We have presented an application of the Lyapunov functional approach for continuous systems, as well as the related definitions and theorems. We have also briefly summarized the results of Koiter extending the Lagrange potential energy criterion to continuous systems.

References

[1] H. ZIEGLER: Some Developments in the Theory of Stability. *Canadian Congress of Applied Mechanics*, 1967.

[2] H. ZIEGLER: *Principles of Structural Stability*. Blaisdell Publishing Co., Waltham, Mass., 1967.

[3] W. T. KOITER: Purpose and Achievements of Research in Elastic Stability. *Proceedings—4th Technical Conference of the Society for Engineering Science*, North Carolina State University, Rayleigh, 1966.

[4] W. T. KOITER: On the Concept of Stability of Equilibrium for Continuous Bodies. *Proceedings—Koninkl. Nederl. Akademie van Wetenschappen*, Series B, Vol. 66, 1963, p. 173.

[5] W. T. KOITER: The Energy Criterion of Stability for Continuous Elastic Bodies. I, II. *Proceedings—Koninkl. Nederl. Akademie van Wetenschappen*, Series B, Vol. 68. 1965, p. 178.

[6] J. W. HUTCHINSON and W. T. KOITER: Postbuckling Theory. *Applied Mechanics Reviews*, Vol. 23, No. 12, December 1970, p. 1353.

[7] R. H. PLAUT: A Study of the Dynamic Stability of Continuous Elastic Systems by Liapunov's Direct Method. *University of California, Berkeley* Report No. AM-67-3. May 1967.

[8] G. H. HARDY, J. E. LITTLEWOOD and G. PÓLYA: *Inequalities*. Cambridge University Press, Cambridge, 1952.

[9] R. J. KNOPS and E. W. WILKES: On Movchan's Theorems for Stability of Continuous Systems. *International Journal of Engineering Science*, Vol. 4, 1966, pp. 303-329.

[10] R. T. SHIELD: On the Stability of Linear Continuous Systems. *Zeitschrift für ange wandte Mathematik und Physik*, Vol. 16, No. 5, May 1965, p. 649.

[11] A. A. MOVCHAN: The Direct Method of Liapunov in Stability Problems of Elastic Systems. *Prikladnaya Matematika i Mekhanika*, Vol. 23, 1959, pp. 483-493; also *Applied Mathematics and Mechanics*, Vol. 23, No. 3, 1959, pp. 686-700.

[12] A. A. MOVCHAN: Stability of Processes with Respect to Two Metrics. *Prikladnaya Matematika i Mekhanika*, Vol. 24, 1960, pp. 988-1001; also *Applied Mathematics and Mechanics*, Vol. 24, No. 6, 1960, pp. 1506-1524.

[13] P. C. PARKS: A Stability Criterion for Panel Flutter Problem via the Second Method of Liapunov. *AIAA Journal*, Vol. 4, No. 1, 1966, pp. 175-178.

[14] P. K. C. WANG: Stability Analysis of a Simplified Flexible Vehicle via Lyapunov's Direct Method. *AIAA Journal*, Vol. 3, No. 9, 1967, pp. 1764-1766.

[15] P. K. C. WANG: Stability Analysis of Elastic and Aeroelastic Systems via Lyapunov's Direct Method. *Journal of the Franklin Institute*, Vol. 281, No. 1, 1966, pp. 51-72.

[16] G. A. HEGEMIER: Stability of Cylindrical Shells Under Moving Loads by the Direct Method of Liapunov. *Journal of Applied Mechanics*, Vol. 34, No. 4, December 1967, p. 991.

[17] A. J. BERGER and L. LAPIDUS: An Introduction to the Stability of Distributed Systems via a Liapunov Functional. *AIChE Journal*, Vol. 14, No. 4, July 1968, p. 558.

[18] V. I. ZUBOV: *Methods of A. M. Lyapunov and their Application*. Leningrad, 1957; English translation, P. Noordhoff Ltd., Groningen, Holland, 1964.

[19] R. H. PLAUT: Displacement Bounds for Beam-Columns With Initial Curvature Subjected to Transient Loads. *International Journal of Solids and Structures*, Vol. 7, No. 9, September 1971, p. 1229.

[20] S. M. HOLZER: Response Bounds for Columns With Transient Loads. *Journal of Applied Mechanics*, Vol. 38, No. 1, March 1971, p. 157.

[21] S. M. HOLZER: Stability and Boundedness Via Liapunov's Direct Method. *Journal of the Engineering Mechanics Division*, ASCE, Vol. 98, No. EM5, October 1972, p. 1273.

5

Some results from nonlinear elasticity theory

We wish to present here some results from the nonlinear theory of elasticity, particularly for large deformations of elastic columns and plates. Complete discussions of nonlinear elasticity theory may be found in the works of Novozhilov [1], and Green and Adkins [2], while briefer discussions are given by Fung [3] and Dym and Shames [4]. The principal complexity in dealing with large deformations is of course the nonlinearity of the equations. Another major problem is due to the fact that, in the general theory, the local stresses are difficult to identify because the deformations cause changes in the areas with respect to which the stresses are measured. Thus, it is of paramount importance to specify in the general case whether one is operating in terms of the original undeformed coordinates, or the deformed set of coordinates that move with the deforming body.

However, for the practical problems of elastic stability, such precision is not required. All that we require is the ability to allow for multiple solutions — remember that uniqueness is a principal feature of classical elasticity — and sufficiently large nonlinearity to allow projections of planar forces to have a significant component out of their initial plane of action.

In this chapter we shall devote ourselves to the development of three useful theories, i.e., large deflections of columns and large deflections of plates and of arches. The theories are fundamentally different, as the first is based on the incompressible (inextensible) behavior of the column centerline, while the last two are based on the coupling between the out-of-plane bending deflection and the (nonlinear) strains in the middle surface of the structure.

5.1   Large bending deflections of columns

From the nonlinear theory of elasticity, we may use the strains ε_{ij} to find the deformed length $(\mathrm{d}s^*)$ of a line element originally of length $(\mathrm{d}s)$ as [4]

$$(\mathrm{d}s^*)^2 = (\mathrm{d}s)^2 + 2\varepsilon_{ij}\,\mathrm{d}x_i\,\mathrm{d}x_j \tag{5.1}$$

where

$$(\mathrm{d}s)^2 = \mathrm{d}x_i\,\mathrm{d}x_i \tag{5.2}$$

and

$$2\varepsilon_{ij} = \frac{\partial u_i}{\partial x_j} + \frac{\partial u_j}{\partial x_i} + \frac{\partial u_k}{\partial x_i}\,\frac{\partial u_k}{\partial x_j} \tag{5.3}$$

In writing down these results in this form, we are using a Lagrangian approach as we are identifying in equation (5.3) the displacement components u_i as functions of the undeformed coordinates x_j.

If we let $u_1 = u(x)$ and $u_3 = w(x)$ be, respectively, axial and transverse displacements measured from the undeformed axis of the column, which lies upon the x-axis (see figure 1), then we can easily calculate the length of an element $(\mathrm{d}l')$ that originally was of length $\mathrm{d}x$, i.e.,

$$(\mathrm{d}l')^2 = (\mathrm{d}x)^2(1 + 2\varepsilon_{xx}) \tag{5.4}$$

where

$$\varepsilon_{xx} = \frac{\mathrm{d}u}{\mathrm{d}x} + \frac{1}{2}\left(\frac{\mathrm{d}u}{\mathrm{d}x}\right)^2 + \frac{1}{2}\left(\frac{\mathrm{d}w}{\mathrm{d}x}\right)^2 \tag{5.5}$$

If the column is *incompressible* during its bending from the straight line configuration, then $\mathrm{d}l' = \mathrm{d}x$, and we must require

$$\left(1 + \frac{\mathrm{d}u}{\mathrm{d}x}\right)^2 + \left(\frac{\mathrm{d}w}{\mathrm{d}x}\right)^2 = 1 \tag{5.6}$$

From figure 1 we can see that

$$\sin\theta = \frac{(\mathrm{d}w/\mathrm{d}x)\,\mathrm{d}x}{AC} = \frac{(\mathrm{d}w/\mathrm{d}x)\,\mathrm{d}x}{\left[\left(1 + \dfrac{\mathrm{d}u}{\mathrm{d}x}\right)^2 + \left(\dfrac{\mathrm{d}w}{\mathrm{d}x}\right)^2\right]^{\frac{1}{2}}(\mathrm{d}x)}$$

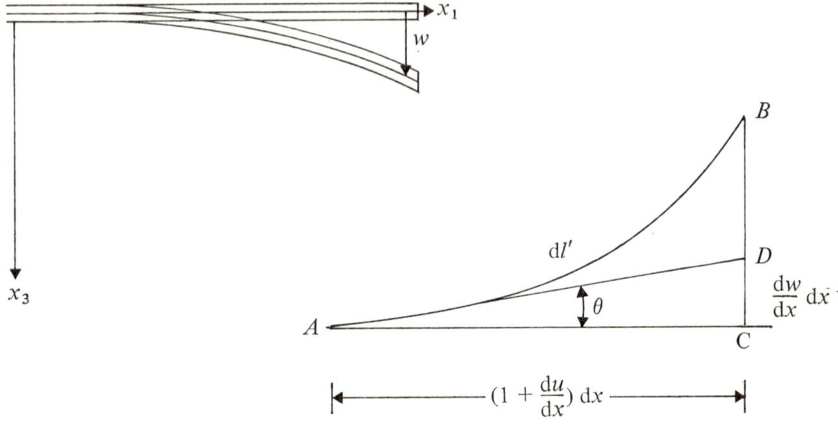

Fig 1. Column geometry.

In view of the condition of incompressibility we can write

$$\sin \theta = \frac{dw}{dx} \tag{5.7}$$

from which it follows by differentiation that

$$\cos \theta \, \frac{d\theta}{dx} = \frac{d^2 w}{dx^2} \tag{5.8}$$

The true curvature of the deformed curve is [1] $1/\rho = d\theta/dl' = d\theta/dx$, so that

$$\frac{1}{\rho} = \frac{d^2 w/dx^2}{\cos \theta} = \frac{w''}{\sqrt{1 - (w')^2}} \tag{5.9}$$

---

[1] We may also obtain this result by considering the deformed coordinates $x_1' = x + u$, $x_3' = w$, and noting a useful result from differential geometry [5]

$$\frac{1}{\rho} = \pm \frac{\dfrac{dx_1'}{dx} \dfrac{d^2 x_3'}{dx^2} - \dfrac{dx_3'}{dx} \dfrac{d^2 x_1'}{dx^2}}{\left\{ \left(\dfrac{dx_1'}{dx} \right)^2 + \left(\dfrac{dx_3'}{dx} \right)^2 \right\}^{3/2}}$$

If we let $x_1' = x$, $x_3' = y(x)$, we could also obtain thusly another familiar result.

where the prime denotes ordinary differentiation with respect to x, and we have also used equation (5.7).

To calculate the potential energy in a bent column we shall need not only the curvature expression (5.9) to calculate the strain energy; we shall also need the shortening of the column due to the bending deflection, so that we can calculate the energy input by the applied axial load. The total shortening would simply be

$$\int_0^L \frac{du}{dx}\, dx = \int_0^L [(1 - (w')^2)^{\frac{1}{2}} - 1]\, dx \tag{5.10}$$

by virtue of the incompressibility condition.

For the familiar elementary theory we would assume the column deformation to be small enough so that $w'(x) < 1$, in which case the curvature simplifies to

$$\frac{1}{\rho} \simeq \frac{d^2 w}{dx^2} \tag{5.11}$$

while the shortening is

$$\int_0^L \frac{du}{dx}\, dx \cong -\frac{1}{2} \int_0^L (w')^2\, dx \tag{5.12}$$

Now the strain energy due to bending, for this simple one-dimensional problem, is

$$U = \frac{E}{2} \int_V [\varepsilon_{xx}^{(z)}]^2\, dV = \frac{E}{2} \int_V \left(\frac{z}{\rho}\right)^2 dV$$

$$U = \frac{EI}{2} \int_0^L \left(\frac{1}{\rho}\right)^2 dx$$

where we have noted the Euler-Bernoulli hypothesis for the bending of beams, in the form

$$\varepsilon_{xx}^{(z)} = \frac{z}{\rho}$$

This equation reflects the fact that the beam deformation is due to bending, and that the neutral axis ($z=0$) is incompressible.

The potential of the applied compressive axial load P is

$$V_P = P[u(L) - u(0)] = P\int_0^L \frac{du}{dx}\,dx$$

which is thus simply the product of the load multiplied by the shortening. The total potential energy is thus

$$\pi = \frac{EI}{2}\int_0^L \left(\frac{1}{\rho}\right)^2 dx + P\int_0^L \frac{du}{dx}\,dx \tag{5.13}$$

5.2   The von Kármán theory of plates

In this section we shall derive the von Kármán theory of plates, which is a nonlinear theory in the sense that it allows for comparatively large rotations of line elements originally normal to the x, y axes, in the plane of the plate (figure 2). These rotation terms allow projections of the in-plane forces $\overline{N}_v$ and $\overline{N}_{vs}$ to be felt, in the transverse direction, normal to the plane of the plate.

This plate theory is a subset of the general nonlinear theory of elasticity [1, 4]. It is derived assuming that the strains and rotations are both small compared to unity, so that we can ignore changes in geometry in the definition of stress components and in the limits of integration needed for work and energy considerations. We further stipulate that the strains will be smaller than the rotations, in the sense described below.

We introduce the linear strain parameters e_{ij} and the rotation parameters ω_{ij} defined as

$$2e_{ij} = u_{i,j} + u_{j,i}$$
$$2\omega_{ij} = u_{i,j} - u_{j,i} \tag{5.14}$$

The strain-displacement equation (5.3) can then be written as

$$2\varepsilon_{ij} = 2e_{ij} + (e_{ki} + \omega_{ki})(e_{kj} + \omega_{kj}) \tag{5.15}$$

For the simplifications discussed, equation (5.15) reduces to [4]:

$$\varepsilon_{ij} = e_{ij} + \tfrac{1}{2}\omega_{ki}\omega_{kj} \tag{5.16}$$

Finally we use the Kirchhoff assumption [1] that lines normal to the undeformed middle surface remain normal to this surface in the deformed geometry and are unextended after deformation. This means that:

$$u_1(x_1, x_2, x_3) = u(x, y) - z\,\frac{\partial w(x, y)}{\partial x} \tag{5.17a}$$

$$u_2(x_1, x_2, x_3) = v(x, y) - z\,\frac{\partial w(x, y)}{\partial y} \tag{5.17b}$$

$$u_3(x_1, x_2, x_3) = w(x, y) \tag{5.17c}$$

where u, v, and w are the displacement components of the middle surface of the plate. We may now give the strain parameters and rotation parameters as follows:

$$
\begin{aligned}
e_{11} &= \frac{\partial u}{\partial x} - z\,\frac{\partial^2 w}{\partial x^2}
&\qquad
e_{12} &= \frac{1}{2}\left(\frac{\partial u}{\partial y} + \frac{\partial v}{\partial x} - 2z\,\frac{\partial^2 w}{\partial x \partial y} \right) \\[2ex]
e_{22} &= \frac{\partial v}{\partial y} - z\,\frac{\partial^2 w}{\partial y^2}
&\qquad
e_{13} &= e_{23} = 0 \\[2ex]
e_{33} &= \frac{\partial w}{\partial z} = 0
\end{aligned}
\tag{5.18}
$$

$$
\begin{aligned}
\omega_{12} &= \frac{1}{2}\left(\frac{\partial u}{\partial y} - \frac{\partial v}{\partial x} \right) \\[2ex]
\omega_{13} &= -\,\frac{\partial w}{\partial x} \\[2ex]
\omega_{23} &= -\,\frac{\partial w}{\partial y}
\end{aligned}
\tag{5.19}
$$

---

[1] The two-dimensional extension of the Euler-Bernoulli hypothesis.

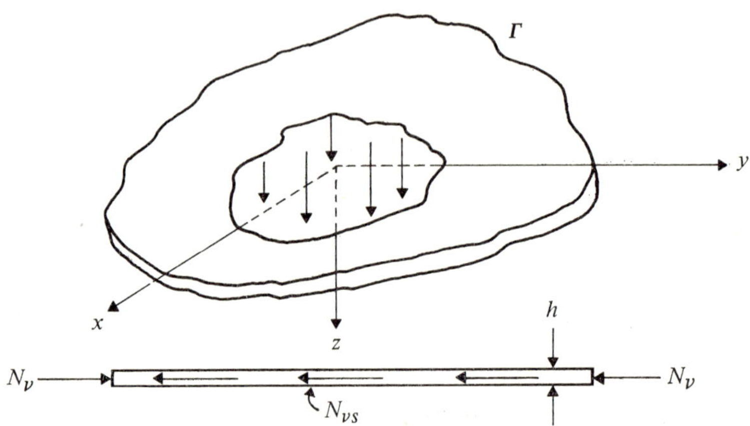

Fig. 2. Loading configuration on a thin plate.

We now observe the following. The rotation parameter ω_{12} approximates a rotation component about the z axis, while ω_{23} and ω_{13} approximate rotation components about axes parallel to the x and y axes, respectively, in the midplane of the plate. For a thin, hence flexible, plate we can reasonably expect that

$$\omega_{12} \ll \omega_{23}, \omega_{13} \tag{5.20}$$

Neglecting ω_{12}, we can now employ equations (5.16, 5.18, 5.19) to find the ε_{ij}:

$$\varepsilon_{11} = \frac{\partial u}{\partial x} + \frac{1}{2}\left(\frac{\partial w}{\partial x}\right)^2 - z\frac{\partial^2 w}{\partial x^2}$$

$$\varepsilon_{22} = \frac{\partial v}{\partial y} + \frac{1}{2}\left(\frac{\partial w}{\partial y}\right)^2 - z\frac{\partial^2 w}{\partial y^2}$$

$$\varepsilon_{33} = \frac{1}{2}\left(\frac{\partial w}{\partial x}\right)^2 + \frac{1}{2}\left(\frac{\partial w}{\partial y}\right)^2$$

$$\varepsilon_{12} = \frac{1}{2}\left(\frac{\partial u}{\partial y} + \frac{\partial v}{\partial x} - 2z\frac{\partial^2 w}{\partial x\partial y}\right) + \frac{1}{2}\frac{\partial w}{\partial x}\frac{\partial w}{\partial y}$$

$$\varepsilon_{13} \cong \varepsilon_{23} \cong 0 \tag{5.21}$$

For a constitutive law we will employ Hooke's law as reduced for plane stress over the thickness of the plate. Thus we shall be concerned here only with ε_{11}, ε_{22}, and ε_{12}. Accordingly, when we consider the total potential energy for the assumptions presented here, we can say

$$\iiint_V \tau_{ij}\delta\varepsilon_{ij}\,\mathrm{d}V$$

$$= \iint_R \int_{-h/2}^{h/2} (\tau_{11}\delta\varepsilon_{11} + 2\tau_{12}\delta\varepsilon_{12} + \tau_{22}\delta\varepsilon_{22})\,\mathrm{d}z\,\mathrm{d}A \tag{5.22}$$

Now we replace the strain terms in the above expression, which we then denote as $\delta^{(1)}U$, the first variation of the plate strain energy,

$$\delta^{(1)}U = \iint_R \int_{-h/2}^{h/2} \left\{ \tau_{11}\left[\frac{\partial\delta u}{\partial x} + \frac{\partial w}{\partial x}\frac{\partial\delta w}{\partial x} - z\frac{\partial^2\delta w}{\partial x^2} \right] \right.$$

$$+ \tau_{12}\left[\frac{\partial\delta u}{\partial y} + \frac{\partial\delta v}{\partial x} - 2z\frac{\partial^2\delta w}{\partial x\partial y} + \frac{\partial w}{\partial x}\frac{\partial\delta w}{\partial y} + \frac{\partial w}{\partial y}\frac{\partial\delta w}{\partial x} \right]$$

$$\left. + \tau_{22}\left[\frac{\partial\delta v}{\partial y} + \frac{\partial w}{\partial y}\frac{\partial\delta w}{\partial y} - z\frac{\partial^2\delta w}{\partial y^2} \right] \right\}\,\mathrm{d}A\,\mathrm{d}z \tag{5.23}$$

Next we integrate with respect to z and introduce resultant stress and moment intensity functions, N_x, N_y and N_{xy}, and M_x, M_y, and M_{xy} respectively:

$$M_x = \int_{-h/2}^{h/2} \tau_{xx}z\,\mathrm{d}z \qquad\qquad N_x = \int_{-h/2}^{h/2} \tau_{xx}\,\mathrm{d}z$$

$$M_y = \int_{-h/2}^{h/2} \tau_{yy}z\,\mathrm{d}z \qquad\qquad N_y = \int_{-h/2}^{h/2} \tau_{yy}\,\mathrm{d}z$$

$$M_{xy} = \int_{-h/2}^{h/2} \tau_{xy}z\,\mathrm{d}z = M_{yx} \qquad\qquad N_{xy} = \int_{-h/2}^{h/2} \tau_{xy}\,\mathrm{d}z = N_{yx} \tag{5.24}$$

where $N_x(N_y)$ is a force in the x direction (y direction) measured per unit length in the y direction (x direction) and where $N_{xy}(N_{yx})$ is a force in the $x(y)$ direction. Similarly the quantity M_x represents a moment per unit length in the y direction, with its vector in the y direction, etc. (figure 3).

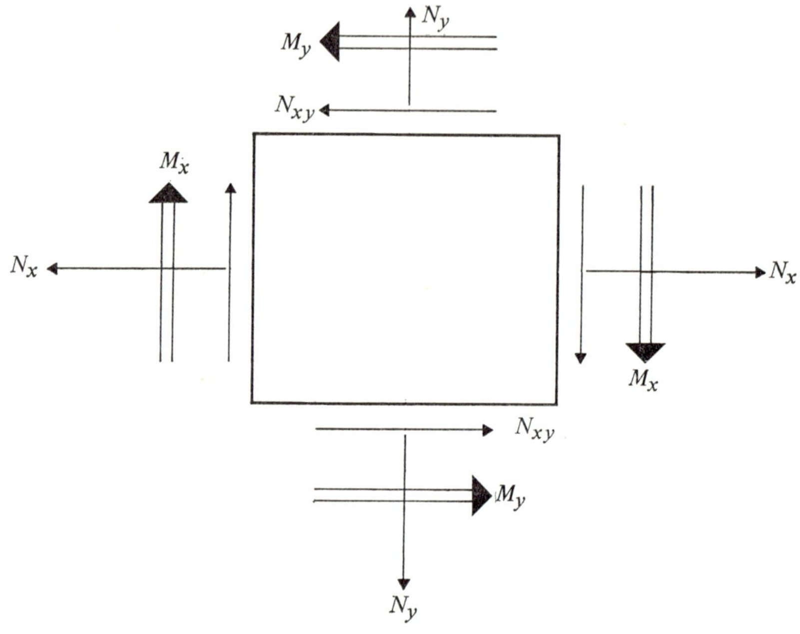

Fig. 3. Stress and moment resultants for a rectangular plate.

These may be (for practical purposes) compared with normal and shear stresses, and we may conclude that $N_{xy} = N_{yx}$. We then get for equation (5.23) the following result:

$$\delta U^{(1)} = \iint_R \left\{ N_x \left(\frac{\partial \delta u}{\partial x} + \frac{\partial w}{\partial x} \frac{\partial \delta w}{\partial x} \right) - M_x \frac{\partial^2 \delta w}{\partial x^2} \right.$$

$$+ N_{xy} \left(\frac{\partial \delta u}{\partial y} + \frac{\partial \delta v}{\partial x} + \frac{\partial w}{\partial x} \frac{\partial \delta w}{\partial y} + \frac{\partial w}{\partial y} \frac{\partial \delta w}{\partial x} \right)$$

$$\left. - 2M_{xy} \frac{\partial^2 \delta w}{\partial x \partial y} + N_y \left(\frac{\partial \delta v}{\partial y} + \frac{\partial w}{\partial y} \frac{\partial \delta w}{\partial y} \right) - M_y \frac{\partial^2 \delta w}{\partial y^2} \right\} \, dx \, dy \quad (5.25)$$

The first variation of the potential of the applied forces takes the form (noting that $\overline{N}_v$ is taken as positive in compression as shown in figure 2)

$$\delta^{(1)}V = -\iint_R q\delta w \,\mathrm{d}x\,\mathrm{d}y + \int_\Gamma \overline{N}_v \delta u_v \,\mathrm{d}s + \int_\Gamma \overline{N}_{vs}\delta u_s \,\mathrm{d}s \qquad (5.26)$$

where u_v and u_s are the in-plane displacements of the boundary of the plate in directions normal and tangential, respectively, to the boundary. We are using the undeformed geometry for the applied loads above rather than the deformed geometry—thereby restricting the result to reasonably small deformations. By using the above result for $\delta^{(1)}V$ and using equation (5.25) for $\delta^{(1)}U$ we may form $\delta^{(1)}\pi$. The total potential energy so formed then approximates the actual total potential energy for the kind of deformation restrictions as are embodied in Kirchhoff's assumptions. And, since we have used undeformed geometry for stresses and external loads, we are limited to small deformations as far as employing this functional. Finally, because we employed equation (5.16) for strain, we are assuming that strains are much smaller than rotations. Thus we have for the total potential energy variation:

$$\delta^{(1)}\pi = 0 = \iint_R \left\{ N_x \left(\frac{\partial \delta u}{\partial x} + \frac{\partial w}{\partial x}\frac{\partial \delta w}{\partial x} \right) - M_x \frac{\partial^2 \delta w}{\partial x^2} \right.$$

$$+ N_{xy} \left(\frac{\partial \delta u}{\partial y} + \frac{\partial \delta v}{\partial x} + \frac{\partial w}{\partial x}\frac{\partial \delta w}{\partial y} + \frac{\partial w}{\partial y}\frac{\partial \delta w}{\partial x} \right)$$

$$\left. - 2M_{xy} \frac{\partial^2 \delta w}{\partial x \partial y} + N_y \left(\frac{\partial \delta v}{\partial y} + \frac{\partial w}{\partial y}\frac{\partial \delta w}{\partial y} \right) - M_y \frac{\partial^2 \delta w}{\partial y^2} \right\} \mathrm{d}x\,\mathrm{d}y$$

$$- \iint_R q\delta w \,\mathrm{d}x\,\mathrm{d}y + \int_\Gamma \overline{N}_v \delta u_v \,\mathrm{d}s + \int_\Gamma \overline{N}_{vs}\delta u_s \,\mathrm{d}s \qquad (5.27)$$

We may proceed to carry out the extremization process. We employ Green's theorem one or more times to get the δu's and δv's out from the partial derivatives. Then we proceed to simplify the expressions in the line integrals by noting from equilibrium (see figure 4) that

$$N_v = N_x a_{vx}^2 + 2N_{xy}a_{vx}a_{vy} + N_y a_{vy}^2 \qquad (5.28a)$$

$$N_{vs} = (N_y - N_x)a_{vx}a_{vy} + N_{xy}(a_{vx}^2 - a_{vy}^2) \qquad (5.28b)$$

where a_{vx} and a_{vy} are the direction cosines of the outward normal of the

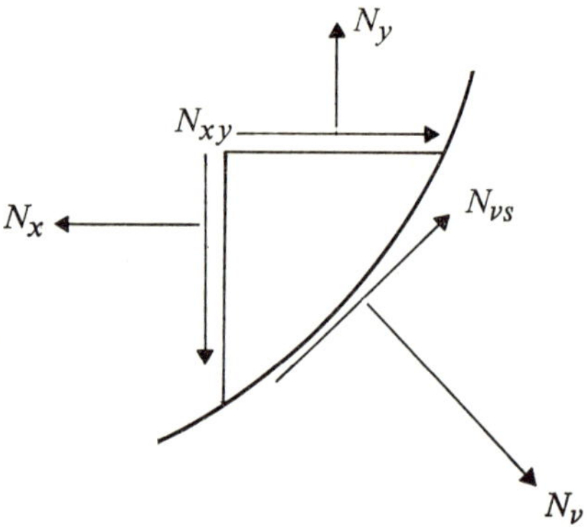

Fig. 4. Boundary loading on a plate segment.

boundary. Furthermore, simple vector projections permit us to say

$$u_v = a_{vx}u + a_{vy}v$$

$$u_s = - a_{vy}u + a_{vx}v \tag{5.29}$$

And we also note that

$$\frac{\partial}{\partial x} = a_{vx}\frac{\partial}{\partial v} - a_{vy}\frac{\partial}{\partial s}$$

$$\frac{\partial}{\partial y} = a_{vy}\frac{\partial}{\partial v} + a_{vx}\frac{\partial}{\partial s} \tag{5.30}$$

Finally we introduce the transverse shear forces of plate theory,

$$Q_v = Q_x a_{vx} + Q_y a_{vy}$$

$$Q_x = \frac{\partial M_x}{\partial x} + \frac{\partial M_{xy}}{\partial y}$$

$$Q_y = \frac{\partial M_y}{\partial y} + \frac{\partial M_{xy}}{\partial x} \tag{5.31}$$

Now utilizing equations (5.28-5.31) we may rewrite equation (5.27) as

$$
\delta^{(1)}\pi = -\iint_R \left\{ \left(\frac{\partial N_x}{\partial x} + \frac{\partial N_{xy}}{\partial y} \right)\delta u + \left(\frac{\partial N_{xy}}{\partial x} + \frac{\partial N_y}{\partial y} \right)\delta v \right.
$$

$$
+ \left[\frac{\partial^2 M_x}{\partial x^2} + 2\frac{\partial^2 M_{xy}}{\partial x \partial y} + \frac{\partial^2 M_y}{\partial y^2} + \frac{\partial}{\partial x}\left(N_x \frac{\partial w}{\partial x} \right) + \frac{\partial}{\partial y}\left(N_{xy}\frac{\partial w}{\partial x} \right) \right.
$$

$$
\left. \left. + \frac{\partial}{\partial x}\left(N_{xy}\frac{\partial w}{\partial y} \right) + \frac{\partial}{\partial y}\left(N_y \frac{\partial w}{\partial y} \right) + q \right]\delta w \right\}\, dx\, dy
$$

$$
+ \int_\Gamma (N_v + \bar{N}_v)\delta u_v\, ds + \int_\Gamma (N_{vs} + \bar{N}_{vs})\delta u_s\, ds - \int_\Gamma M_v \frac{\partial \delta w}{\partial v}\, ds
$$

$$
+ \int_\Gamma \left(Q_v + \frac{\partial M_{vs}}{\partial s} + N_v \frac{\partial w}{\partial v} + N_{vs}\frac{\partial w}{\partial s} \right)\delta w\, ds - [M_{vs}\delta w]_\Gamma = 0
$$

$$
(5.32)
$$

The last expression accounts for 'corners' in the boundary. From the above equations we may now make a series of deductions. First in the region R we can conclude that:

$$
\frac{\partial N_x}{\partial x} + \frac{\partial N_{xy}}{\partial y} = 0 \tag{5.33a}
$$

$$
\frac{\partial N_{xy}}{\partial x} + \frac{\partial N_y}{\partial y} = 0 \tag{5.33b}
$$

$$
\frac{\partial^2 M_x}{\partial x^2} + 2\frac{\partial^2 M_{xy}}{\partial x \partial y} + \frac{\partial^2 M_y}{\partial y^2} + \frac{\partial}{\partial x}\left(N_x \frac{\partial w}{\partial x} \right) + \frac{\partial}{\partial y}\left(N_{xy}\frac{\partial w}{\partial x} \right)
$$

$$
+ \frac{\partial}{\partial x}\left(N_{xy}\frac{\partial w}{\partial y} \right) + \frac{\partial}{\partial y}\left(N_y \frac{\partial w}{\partial y} \right) + q = 0 \tag{5.33c}
$$

The first two equations, above, clearly are identical to the equations of equilibrium for plane stress, as is to be expected. We shall use these equations now to simplify the third equation after we carry out the differentiation operation on the expressions involving products. We are thus able to eliminate in this way expressions involving the first partial of w. We get

$$\frac{\partial^2 M_x}{\partial x^2} + 2\frac{\partial^2 M_{xy}}{\partial x \partial y} + \frac{\partial^2 M_y}{\partial y^2} + N_x\frac{\partial^2 w}{\partial x^2} + 2N_{xy}\frac{\partial^2 w}{\partial x \partial y}$$

$$+ N_y\frac{\partial^2 w}{\partial y^2} + q = 0 \tag{5.34}$$

Now, comparing equation (5.34) with the classical case we see that we have here introduced *nonlinear terms* $[N_x(\partial^2 w/x^2) + 2N_{xy}(\partial^2 w/\partial x \partial y) + N_y(\partial^2 w/\partial y^2)]$ involving the in-plane force intensities as additional 'transverse loading' terms. [1]

Considering next the remainder of equation (5.32) we can stipulate the following boundary conditions along Γ:

$$\text{Either } N_v = -\,\overline{N}_v \quad \text{or} \quad u_v \text{ is specified} \tag{5.35a}$$

$$\text{Either } N_{vs} = -\,\overline{N}_{vs} \quad \text{or} \quad u_s \text{ is specified} \tag{5.35b}$$

$$\text{Either } M_v = 0 \quad \text{or} \quad \frac{\partial w}{\partial v} \text{ is specified} \tag{5.35c}$$

$$\text{Either } Q_v + \frac{\partial M_{vs}}{\partial s} + N_v\frac{\partial w}{\partial v} + N_{vs}\frac{\partial w}{\partial s} = 0 \quad \text{or} \quad w \text{ is specified}$$

$$\tag{5.35d}$$

$$\text{At discontinuities } [M_{vs}\delta w] = 0 \tag{5.35e}$$

The last three conditions are familiar from work on plates except that the effective shear force $(Q_v + \partial M_{vs}/\partial s)$ is now augmented by projections of the in-plane plate forces at the plate edges.

The equations of equilibrium may be 'solved' if a constitutive law is used. We will employ here the familiar Hooke's law for plane stress. We will use the constitutive law to replace the resultant intensity functions by appropriate derivatives of the displacement field of the midplane of the plate. Consider for example the quantity M_x. We have, using Hooke's law,

$$M_x = \int_{-h/2}^{h/2} z\tau_{xx}\,\mathrm{d}z = \int_{-h/2}^{h/2} z\left(\frac{E}{1-v^2}\right)(\varepsilon_{xx} + v\varepsilon_{yy})\,\mathrm{d}z$$

---

[1] In the Russian literature these terms are often referred to as the 'reduced loads' of the in-plane forces or the 'reduced forces'.

$$= \int_{-h/2}^{h/2} z \frac{E}{1 - v^2} \left[\frac{\partial u}{\partial x} + \frac{1}{2} \left(\frac{\partial w}{\partial x} \right)^2 - z \frac{\partial^2 w}{\partial x^2} + v \frac{\partial v}{\partial y} \right.$$

$$\left. + \frac{v}{2} \left(\frac{\partial w}{\partial y} \right)^2 - vz \frac{\partial^2 w}{\partial y^2} \right] dz$$

Integrating and inserting limits we get

$$M_x = \left(\frac{E}{1 - v^2} \right) \left(\frac{h^3}{12} \right) \left(-\frac{\partial^2 w}{\partial x^2} - v \frac{\partial^2 w}{\partial y^2} \right)$$

$$= -D \left[\frac{\partial^2 w}{\partial x^2} + v \frac{\partial^2 w}{\partial y^2} \right] \tag{5.36}$$

where D is the familiar bending rigidity, given by $D = Eh^3/12(1 - v^2)$. Similarly we have

$$M_y = - \left[\frac{\partial^2 w}{\partial y^2} + v \frac{\partial^2 w}{\partial x^2} \right] \tag{5.37a}$$

$$M_{xy} = - (1 - v) D \frac{\partial^2 w}{\partial x \partial y} \tag{5.37b}$$

$$N_x = C \left\{ \left[\frac{\partial u}{\partial x} + \frac{1}{2} \left(\frac{\partial w}{\partial x} \right)^2 \right] + v \left[\frac{\partial v}{\partial y} + \frac{1}{2} \left(\frac{\partial w}{\partial y} \right)^2 \right] \right\} \tag{5.37c}$$

$$N_y = C \left\{ \left[\frac{\partial v}{\partial y} + \frac{1}{2} \left(\frac{\partial w}{\partial y} \right)^2 \right] + v \left[\frac{\partial u}{\partial x} + \frac{1}{2} \left(\frac{\partial w}{\partial x} \right)^2 \right] \right\} \tag{5.37d}$$

$$N_{xy} = C \left(\frac{1 - v}{2} \right) \left[\frac{\partial u}{\partial y} + \frac{\partial v}{\partial x} + \left(\frac{\partial w}{\partial x} \right) \left(\frac{\partial w}{\partial y} \right) \right] \tag{5.37e}$$

where C is the extensional rigidity,

$$C = \frac{Eh}{1 - v^2} \tag{5.38}$$

We could now substitute for the resultant intensity functions using the above relations to get the equilibrium equations in terms of the displacement components of the midplane plate. However, we shall follow another route which leads to a somewhat less complicated system of equations.

Note accordingly that equations (5.33a, b) will be individually satisfied if we define an Airy stress function,

$$N_x = \frac{\partial^2 F}{\partial y^2} \qquad N_y = \frac{\partial^2 F}{\partial x^2} \qquad N_{xy} = -\frac{\partial^2 F}{\partial x \partial y} \qquad (5.39)$$

Then, replacing M_x, M_y, and M_{xy} it is a simple matter to show that the last of equations (5.33) can be written

$$D\nabla^4 w = \frac{\partial^2 F}{\partial y^2}\frac{\partial^2 w}{\partial x^2} - 2\frac{\partial^2 F}{\partial x \partial y}\frac{\partial^2 w}{\partial x \partial y} + \frac{\partial^2 F}{\partial x^2}\frac{\partial^2 w}{\partial y^2} + q \qquad (5.40)$$

We now have a single partial differential equation with two dependent variables, w and F. Since we are now studying in-plane effects using a stress approach we must assure the compatibility of the in-plane displacements. This will give us a second companion equation to go with the above equation. To do this, we shall seek to relate the strains ε_{xx}, ε_{yy}, and ε_{xy} at the midplane surface in such a way that when we employ equations (5.21) to replace the strains we end up with a result that does not contain the in-plane displacement components u and v. Thus you may readily demonstrate that substituting equations (5.21) into the expression

$$\left[\frac{\partial^2 \varepsilon_{xx}}{\partial y^2} + \frac{\partial^2 \varepsilon_{yy}}{\partial x^2} - 2\frac{\partial^2 \varepsilon_{xy}}{\partial x \partial y}\right]_{z=0} \text{gives}$$

$$\left(\frac{\partial^2 w}{\partial x \partial y}\right)^2 - \left(\frac{\partial^2 w}{\partial x^2}\right)\left(\frac{\partial^2 w}{\partial y^2}\right).$$

That is,

$$\left[\frac{\partial^2 \varepsilon_{xx}}{\partial y^2} + \frac{\partial^2 \varepsilon_{yy}}{\partial x^2} - 2\frac{\partial^2 \varepsilon_{xy}}{\partial x \partial y}\right]_{z=0} = \left(\frac{\partial^2 w}{\partial x \partial y}\right)^2 - \left(\frac{\partial^2 w}{\partial x^2}\right)\left(\frac{\partial^2 w}{\partial y^2}\right)$$

Since the above equation ensures the proper relation of strains at the mid-plane surface to the midplane displacement component w without explicitly involving in-plane displacement components u and v, it serves as the desired compatibility equation for the strains at the midplane surface. We next express the compatibility equation in terms of the stress resultant intensity function, using the plane stress Hooke's law,

$$\frac{1}{Eh}\left[\frac{\partial^2 N_x}{\partial y^2} - v\,\frac{\partial^2 N_y}{\partial y^2} + \frac{\partial^2 N_y}{\partial x^2} - v\,\frac{\partial^2 N_x}{\partial x^2} + 2(1+v)\,\frac{\partial^2 N_{xy}}{\partial x \partial y}\right]$$

$$= \left(\frac{\partial^2 w}{\partial x \partial y}\right)^2 - \left(\frac{\partial^2 w}{\partial x^2}\right)\left(\frac{\partial^2 w}{\partial y^2}\right)$$

and then using the stress function,

$$\nabla^4 F = Eh\left[\left(\frac{\partial^2 w}{\partial x \partial y}\right)^2 - \left(\frac{\partial^2 w}{\partial x^2}\right)\left(\frac{\partial^2 w}{\partial y^2}\right)\right] \tag{5.41}$$

The above equation and equation (5.40), which we now rewrite,

$$D\nabla^4 w = \frac{\partial^2 F}{\partial y^2}\,\frac{\partial^2 w}{\partial x^2} - 2\,\frac{\partial^2 F}{\partial x \partial y}\,\frac{\partial^2 w}{\partial x \partial y} + \frac{\partial^2 F}{\partial x^2}\,\frac{\partial^2 w}{\partial y^2} + q \tag{5.42}$$

are the celebrated *von Kármán plate equations*. Note that they are still highly nonlinear. The equations, furthermore, have considerable mutual symmetry. This is brought out by defining the nonlinear operator L as follows:

$$L(p, q) = \frac{\partial^2 p}{\partial y^2}\,\frac{\partial^2 q}{\partial x^2} - 2\,\frac{\partial^2 p}{\partial x \partial y}\,\frac{\partial^2 q}{\partial x \partial y} + \frac{\partial^2 p}{\partial x^2}\,\frac{\partial^2 q}{\partial y^2} \tag{5.43}$$

Then the von Kármán plate equations can be given as follows:

$$\nabla^4 F = -\frac{Eh}{2}\,L(w, w) \tag{5.44a}$$

$$D\nabla^4 w = L(F, w) + q \tag{5.44b}$$

The deletion of the nonlinear operator leaves us with the (uncoupled) plane stress problem of two-dimensional elasticity theory, as well as the classic plate bending equation.

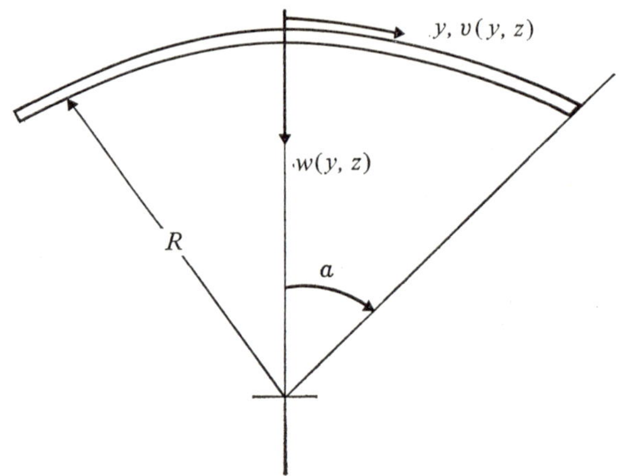

Fig. 5. Arch geometry.

5.3   Some arch equations

The curved arch of uniform radius is a structure where the bending deflection strongly couples with the strain in the middle surface. However, in terms of many of the results obtained, it is not especially sensitive to the compressibility of its centerline. Thus it appears to have behavior that is characteristic both of the column and of the plate.

The equations for an arch are derived here from the nonlinear strain-displacement equations of elasticity, given in terms of an orthogonal curvilinear coordinate system. Specialized to one-dimension, for the coordinate system shown in figure 5, we have [6]

$$\varepsilon_{yy} = \frac{\partial v}{\partial y} - \frac{w}{R} + \frac{1}{2}\left(\frac{\partial v}{\partial y} - \frac{w}{R}\right)^2 + \frac{1}{2}\left(\frac{\partial w}{\partial y} + \frac{v}{R}\right)^2 \tag{5.44}$$

where $v(y, z)$ is the tangential displacement and $w(y, z)$ is the radial displacement. To draw an analogy with our plate theory development, it is not difficult to see that equation (5.44) is of the form

$$\varepsilon_{yy} = e_{yy} + \tfrac{1}{2}e_{yy}^2 + \tfrac{1}{2}\omega_{yz}^2 \tag{5.45}$$

Under a set of assumptions similar to those used before, we would argue
that our strain would be sufficiently accurately expressed as:

$$\varepsilon_{yy} = e_{yy} + \tfrac{1}{2}\omega_{yz}^2 \tag{5.46}$$

or

$$\varepsilon_{yy} = \frac{\partial v}{\partial y} - \frac{w}{R} + \frac{1}{2}\left(\frac{\partial w}{\partial y} + \frac{v}{R}\right)^2 \tag{5.47}$$

In analogy to the Kirchoff plate assumptions, for the details of the displacement distribution through the thickness, we assume here that

$$v(y, z) = (1 - z/R)\, v\,(y) - z\,\frac{dw}{dy} = v - z\left(\frac{dw}{dy} + \frac{v}{R}\right)$$

$$w(y, z) = w(y) \tag{5.48}$$

Then upon substitution of equations (5.48) into equation (5.47), and considering the arch to be thin, we would find that

$$\varepsilon_{yy} = \frac{dv}{dy} - \frac{w}{R} + \frac{1}{2}\left(\frac{dw}{dy} + \frac{v}{R}\right)^2 - z\left(\frac{d^2w}{dy^2} + \frac{1}{R}\frac{dv}{dy}\right) \tag{5.49}$$

This will be our basic kinematic equation for the development of our arch
theories. We note in passing that equation (5.49) is a one-dimensional subset
of the nonlinear shell equations of Sanders [7], and is also referred to as a
nonlinear ring equation [8].

We may then define stress and moment resultants,

$$N = \int_{-h/2}^{h/2} b\sigma\, dz, \quad M = \int_{-h/2}^{h/2} b\sigma z\, dz \tag{5.50}$$

which then yields, in conjunction with the one-dimensional stress-strain law,
$\sigma = E\varepsilon_{yy}$,

$$N = EA\left[\frac{dv}{dy} - \frac{w}{R} + \frac{1}{2}\left(\frac{dw}{dy} + \frac{v}{R}\right)^2\right]$$

$$M = -EI\left(\frac{d^2w}{dy^2} + \frac{1}{R}\frac{dv}{dy}\right) \tag{5.51}$$

The first variation of the strain energy can then be deduced to be

$$\delta^{(1)}U_e = \int_{-\alpha R}^{\alpha R}\left\{ N\left[\frac{\mathrm{d}\delta v}{\mathrm{d}y} - \frac{\delta w}{R} + \left(\frac{\mathrm{d}w}{\mathrm{d}y} + \frac{v}{R}\right)\left(\frac{\mathrm{d}\delta w}{\mathrm{d}y} + \frac{\delta v}{R}\right)\right]\right.$$
$$\left. - M\left(\frac{\mathrm{d}^2\delta w}{\mathrm{d}y^2} + \frac{1}{R}\,\frac{\mathrm{d}\delta v}{\mathrm{d}y}\right)\right\}\mathrm{d}y \tag{5.52}$$

For the potential of the applied loading, we will consider only a uniform, radial pressure, so that

$$\delta^{(1)}\pi = \delta^{(1)}U_e - \int_{-\alpha R}^{\alpha R} q\,\delta w\,\mathrm{d}y \tag{5.53}$$

which yields the equations of equilibrium

$$\frac{\mathrm{d}N}{\mathrm{d}y} - \frac{N}{R}\left(\frac{\mathrm{d}w}{\mathrm{d}y} + \frac{v}{R}\right) - \frac{1}{R}\,\frac{\mathrm{d}M}{\mathrm{d}y} = 0$$
$$\frac{N}{R} + \frac{\mathrm{d}}{\mathrm{d}y}\left[N\left(\frac{\mathrm{d}w}{\mathrm{d}y} + \frac{v}{R}\right)\right] + \frac{\mathrm{d}^2 M}{\mathrm{d}y^2} + q = 0 \tag{5.54}$$

We also note in passing that the direction of the applied pressure on an arch or on a ring is of crucial importance, that is, it is possible to load an arch so that the pressure is always in the same direction, so that the pressure is hydrostatic, or so that the pressure is centrally directed [9]. In the first case, which corresponds to our equation (5.53), the loading is such that it is always directed parallel to its original direction, prior to buckling. In the other two cases the direction of the loading changes during deformation, either to continuously remain normal to the local centerline (hydrostatic) or to be continuously directed towards the initial center of curvature. The differences in the loading can result in significantly different buckling behavior [9, 10].

Another important class of arch problems is that of pressure-loaded *shallow arches*. These are arches where the arch rise (the altitude of the crown, over the base) is small compared to the arch span. In such a case we may effectively argue that the rotation ω_{yz} behaves very much like that of a beam, i.e., we should approximate it as

$$\omega_{yz} \cong \frac{\mathrm{d}w}{\mathrm{d}y} \tag{5.55}$$

In this instance our basic kinematic expression becomes

$$\varepsilon_{yy} = \frac{\mathrm{d}v}{\mathrm{d}y} - \frac{w}{R} + \frac{1}{2}\left(\frac{\mathrm{d}w}{\mathrm{d}y}\right)^2 - z\,\frac{\mathrm{d}^2 w}{\mathrm{d}y^2} \tag{5.56}$$

which leads to the equilibrium equations

$$\frac{\mathrm{d}N}{\mathrm{d}y} = 0, \qquad \frac{N}{R} + \frac{\mathrm{d}}{\mathrm{d}y}\left(N\,\frac{\mathrm{d}w}{\mathrm{d}y}\right) + \frac{\mathrm{d}^2 M}{\mathrm{d}y^2} + q = 0 \tag{5.57}$$

Note, for later use, that the first of these equations implies that N is a constant. This simplification is very useful, for it can be used to help obtain an exact, closed form solution for the complete, nonlinear, buckling and postbuckling problem for shallow arches [11].

5.4   Summary

We have presented herein brief derivations of the equations required for the discussion of the buckling and postbuckling behavior of columns, plates and arches.

References

[1] V. V. NOVOZHILOV: *Foundations of the Nonlinear Theory of Elasticity.* Graylock Press, Rochester, N.Y., 1953.

[2] A. E. GREEN and J. E. ADKINS: *Large Elastic Deformations.* Clarendon Press, Oxford, 1970.

[3] Y. C. FUNG: *Foundations of Solid Mechanics.* Prentice Hall, Englewood Cliffs, New Jersey, 1965.

[4] C. L. DYM and I. H. SHAMES: *Solid Mechanics: A Variational Approach.* McGraw-Hill Book Company, New York, 1973.

[5] J. J. STOKER: *Differential Geometry.* Wiley-Interscience, New York, 1969.

[6] H. L. LANGHAAR: *Energy Methods in Applied Mechanics.* John Wiley and Sons, New York, 1962.

[7] J. L. SANDERS: Nonlinear Theories for Thin Shells. *Quarterly of Applied Mathematics,* Vol. 21, 1963, p. 21.

[8] E. CHWALLA and C. F. KOLLBRUNNER: Beitrage sun Knickproblem des Bogentragers und des Rahmen. *Stahlbau,* Vol. 11, 1938, p. 73.

[9] J. SINGER and C. D. BABOCK: On the Buckling of Rings Under Constant Directional and Centrally Directed Pressure. *Journal of Applied Mechanics,* Vol. 37, 1970, p 215.

[10] C. L. DYM: Buckling and Postbuckling Behavior of Steep Compressible Arches. *International Journal of Solids and Structures,* Vol. 9, 1973, p. 129.

[11] H. L. SCHREYER and E. F. MASUR: Buckling of Shallow Arches. *Journal of the Engineering Mechanics Division,* ASCE, Vol. 92, 1966, p. 1.

6

Buckling and postbuckling of elastic columns

6.1 Energy criteria for column buckling

In this section we wish to consider only the buckling problem of the pinned-pinned Euler column. Thus combining the linearized curvature (5.11) and shortening (5.12) relations with the total potential energy functional (5.16), we have

$$\pi = \frac{EI}{2} \int_0^L (w'')^2 \, dx - \frac{P}{2} \int_0^L (w')^2 \, dx \tag{6.1}$$

The vanishing of the first variation of the functional (6.1) leads immediately to the eigenvalue problem defined by the differential equation

$$EI\, w^{IV} + Pw'' = 0 \tag{6.2}$$

and the associated boundary conditions which for the pinned-pinned column take the normal form, i.e.,

$$w(0) = w''(0) = w(L) = w''(L) = 0 \tag{6.3}$$

It is a straightforward matter to show that a solution to the system (6.2, 6.3) exists only if the load P takes on one of the values

$$P = n^2 \frac{\pi^2 EI}{L^2} = n^2 P_E, \quad n = 1, 2, 3 \ldots \tag{6.4}$$

and that the solution corresponding to each of the *eigenvalues* (6.4) is a corresponding *eigenfunction*,

$$w_n(x) = c_n \sin \frac{n\pi x}{L}, \quad n = 1, 2, 3 \ldots \tag{6.5}$$

where the c_n are arbitrary. Thus, a non-trivial solution exists only when P takes on one of the values prescribed by equation (6.4). For all other values of P the solution is the trivial one, $w(x) = 0$, which is to say that for all these other values the column remains straight and undeformed.

This is a *bifurcation* phenomenon, that is, at $P = P_E$ more than one equilibrium path exists. The column can follow only one path.

Thus it doesn't require much to say that our analysis of the buckling of columns has been of the *equilibrium* variety, i.e., we have simply asked whether there exist any bent configurations of the loaded column that are fairly close (remember that we've linearized) to the straight line and that are in equilibrium under the axial load P.

We have indeed found the answer to the equilibrium question, but without a few additional judgements and assumptions we are not likely to find it a very useful guide to the column's stability. Thus we will now take a different approach: we shall apply the minimum energy criterion, i.e., we shall try to find the condition under which the energy reaches a minimum, and remains positive definite about that minimum. This, according to Theorem 4.6, will yield the stable equilibrium state.

We consider the energy variation defined by

$$\delta^{(T)}\pi = \pi(w + \delta w) - \pi(w) \tag{6.6}$$

which takes the form

$$\delta^{(T)}\pi = \delta^{(1)}\pi + \delta^{(2)}\pi \tag{6.7}$$

where

$$\delta^{(1)}\pi = EI \int_0^L w'' \, \delta w'' \, \mathrm{d}x - P \int_0^L w' \, \delta w' \, \mathrm{d}x \tag{6.8}$$

and

$$\delta^{(2)}\pi = \frac{EI}{2} \int_0^L (\delta w'')^2 \, \mathrm{d}x - \frac{P}{2} \int_0^L (\delta w')^2 \, \mathrm{d}x \tag{6.9}$$

Now the vanishing of the first variation corresponds to seeking that point where the minimum of the energy – the equilibrium point – is to be found. This is a direct extension of the vanishing of the first derivative of the potential energy for discrete systems. And here, the vanishing of the first variation led precisely to the eigenvalue problem solved above. In fact, with $w(x)$ given by equation (6.5), and $\delta w(x)$ given as [1]

$$\delta w(x) = (\delta c_n) \sin \frac{n\pi x}{L} \tag{6.10}$$

we can easily show that

$$\delta^{(1)}\pi = \left[EI \left(\frac{n\pi}{L} \right)^4 - P \left(\frac{n\pi}{L} \right)^2 \right] c_n \delta c_n \left(\frac{L}{2} \right) \tag{6.11}$$

Obviously the first variation vanishes whenever $P = n^2 P_E$. Thus, so far, the vanishing of the first variation yields about the same information as the equilibrium analysis.

If we now use the displacement variation (6.10) to calculate the second variation (6.9), we would obtain

$$\delta^{(2)}\pi = \left[EI \left(\frac{n\pi}{L} \right)^4 - P \left(\frac{n\pi}{L} \right)^2 \right] \frac{(\delta c_n)^2}{2} \left(\frac{L}{2} \right) \tag{6.12}$$

For a minimum at the equilibrium state, we must have $\delta^{(2)}\pi$ positive for all permissible loads, for all values of n! Thus, to insure that $\delta^{(2)}\pi > 0$ for all n, we must have

$$P < \frac{\pi^2 EI}{L^2} = P_E \tag{6.13}$$

In other words, *to guarantee a true minimum of the potential energy, we must require that P be less than the Euler buckling load P_E.* By Theorem 4.6, then, it follows that the column will be stable only when $P < P_E$.

---

[1] We should have mentioned that all variations are small, arbitrary, and consistent with all geometric constraints. In calculations such as the one above, it is clearly most satisfying to choose the variation as we have done in equation (6.10).

The Trefftz criterion is based on the positivity of the second variation of the potential energy, although the stability question is asked in a slightly different form. We ask here for the condition for which $\delta^{(2)}\pi$ ceases to be positive definite. From equation (6.12) we could immediately answer that question almost exactly as we have just answered the question of when equation (6.12) remains positive definite. Or, equivalently, we could ask for the values of P for which

$$\delta^{(1)}\left[\delta^{(2)}\pi\right] = 0 \tag{6.14}$$

or, from equation (6.9) with $\eta = \delta w$, we seek P such that

$$\delta^{(1)}\left[\frac{EI}{2}\int_0^L (\eta'')^2 \, \mathrm{d}x - \frac{P}{2}\int_0^L (\eta')^2 \, \mathrm{d}x\right] = 0 \tag{6.15}$$

It is easy to verify that this yields an eigenvalue problem for the second variation that is the same as our eigenvalue problem for $w(x)$ earlier, a 'coincidence' that is due to the quadratic nature of the energy functional (6.1). Thus the Trefftz criterion yields the result *that the second variation ceases to be positive definite when* $P = n^2 P_E$, or, for our purposes, when $P = P_E$.

Thus, in this brief analysis, we have seen that the equilibrium approach yields the information that an adjacent equilibrium state exists for $P = P_E$, while the energy criteria not only yielded the existence of this state but indicated that for $P < P_E$ the straight line configuration was stable. Also, for $P \geq P_E$, stability can no longer be guaranteed.

6.2   Imperfection and kinetic criteria for buckling

In this section we should like to outline very briefly the imperfection and kinetic criteria of buckling, as applied to Euler columns [1, 2, 3]. Again we shall consider only the linearized versions of these problems; we shall, however, return to the imperfect column problem again to treat the non-linear effects as well.

Thus, to do an *imperfection analysis*, let us consider an initially deformed or imperfect column, whose deviation from the straightline configuration is given by $w_0(x)$ (figure 1).

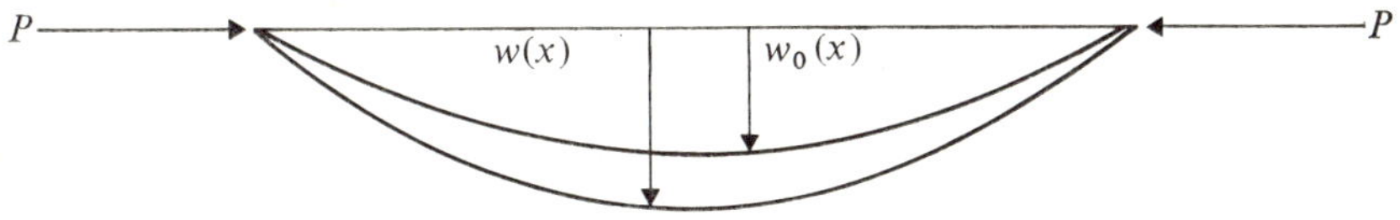

Fig. 1. An imperfect column.

We can easily modify the curvature and shortening expressions to account for the initial deviation. The notation $w(x)$ will still be retained for the total deflection from the straight line configuration, so that $w(x) - w_0(x)$ is the additional deflection due to the axial load.

Since the bending strain energy is proportional to the square of the bending strain (or stress), and since that strain is proportional to the curvature change, it seems perfectly reasonable to take the curvature change as $(w''(x) - w_0''(x))$ so that the column is not stressed by the initial deviation. Further, the work done by the force P is calculated as it acts through the difference in shortening caused by the additional deflection, i.e., through $(du/dx - du_0/dx)$, and where the latter term is given by equation (5.12) with $w(x)$ replaced by $w_0(x)$.

Then the total potential energy for the deformed column is

$$\pi = \frac{EI}{2} \int_0^L (w'' - w_0'')^2 \, dx - \frac{P}{2} \int_0^L [(w')^2 - (w_0')^2] \, dx \qquad (6.16)$$

The vanishing of the first variation leads to the differential equation

$$EI(w^{IV} - w_0^{IV}) + Pw'' = 0 \qquad (6.17)$$

plus the usual boundary conditions. Note that equation (6.17) is inhomogeneous, and thus we no longer have an eigenvalue problem.

If we assume that the (known) initial imperfection can be represented as

$$w_0(x) = \sum_{n=1}^{\infty} a_n \sin \frac{n\pi x}{L} \qquad (6.18)$$

then we can find, for the pinned-pinned column, a solution $w(x)$ in the form

$$w(x) = \sum_{n=1}^{\infty} b_n \sin \frac{n\pi x}{L} \tag{6.19}$$

A straightforward manipulation reveals that

$$b_n = \frac{a_n}{1 - P/n^2 P_E} \tag{6.20}$$

Thus the total deflection is

$$w(x) = \sum_{n=1}^{\infty} \frac{a_n}{1 - (1/n^2)\,(P/P_E)} \sin \frac{n\pi x}{L} \tag{6.21}$$

Note that here, unlike the equilibrium result of the perfect column, a non-trivial deflection exists for loads other than the eigenvalues. As the load increases from zero, say, and approaches the Euler load, the first term in the series blows up. Thus we may take as a stability criterion, with qualifications because of the linearization, the load for which the initial deviations from straightness become amplified beyond any finite bound. Also, it is clear that the only term in the series (6.18, 6.21) that is significant is the first one, no matter how small a_1 is, so long as it is not zero.

For the first term, since we can identify $\bar{\varepsilon} = a_1/L$ and $\varepsilon + \bar{\varepsilon} = b_1/L$ as the maximum values of the deviation and of the total deflection, we can also write

$$\varepsilon + \bar{\varepsilon} = \frac{\bar{\varepsilon}}{1 - P/P_E} \tag{6.22a}$$

or

$$(1 - P/P_E)\varepsilon = (P/P_E)\,\bar{\varepsilon} \tag{6.22b}$$

The first of these forms can be used to develop the famous Southwell plot [2], while the second will serve as the basis for an interesting comparison in the sequel.

To formulate the *kinetic criterion,* we observe that we shall be asking for that value of the axial load for which small oscillations about the straight line configuration cannot be maintained. To examine these small oscillations about the straight line equilibrium configuration, we can modify equation

(6.2) by adding the inertia term as a D'Alembert force, or by using Hamilton's principle [4]. The result of either calculation is the partial differential equation

$$\rho A \, \frac{\partial^2 w}{\partial t^2} + EI \frac{\partial^4 w}{\partial x^4} + P \, \frac{\partial^2 w}{\partial x^2} = 0 \tag{6.23}$$

For free vibrations of the pinned column, we take a solution in the form

$$w(x, t) = \sum_{n=1}^{\infty} A_n \sin \frac{n\pi x}{L} \, e^{i\omega_n t} \tag{6.24}$$

where the frequencies are determined by satisfaction of equation (6.23). They are, then,

$$\omega_n^2 = \left(\frac{1}{\rho A} \right) \left(\frac{n\pi}{L} \right)^2 (n^2 P_E - P) \tag{6.25}$$

Notice from this result that when $P > P_E$, ω_1 becomes imaginary and, as a result, the first term in the expansion (6.24) diverges as time grows large. Thus, for loads in excess of the Euler load, the conclusion is that small oscillations cannot be maintained, and the column is thus considered unstable according to the kinetic criterion, *for the linearized theory.*

6.3   The elastica

In this section we wish to present the very elegant (closed-form) solution to the large displacement version of the column problem. From the previous chapter we know that

$$\frac{1}{\rho} = \frac{d\theta}{dx}, \qquad \frac{du}{dx} = \cos \theta - 1 \tag{6.26}$$

so that the total potential energy is

$$\pi = \frac{EI}{2} \int_0^L \left(\frac{d\theta}{dx} \right)^2 dx + P \int_0^L (\cos \theta - 1) \, dx \tag{6.27}$$

As the column is incompressible, with $dx = dl'$, we may here associate x with the arc length along the column. A straightforward variational procedure yields the Euler-Lagrange equation,

$$EI \frac{d^2\theta}{dx^2} + P \sin \theta = 0 \tag{6.28}$$

The moment-free boundary condition can then be expressed as

$$M = EI \frac{d\theta}{dx} = 0 \tag{6.29}$$

Equation (6.28), although harmless in appearance, is in fact rather difficult to solve because of the nonlinearity inherent in the term $\sin \theta$. However, we can obtain an implicit solution in terms of elliptic integrals. First, multiply equation (6.28) by $d\theta/dx$, so that

$$EI \frac{d\theta}{dx} \frac{d^2\theta}{dx^2} + P \sin \theta \frac{d\theta}{dx} = 0$$

which can be written as

$$\frac{d}{dx} \left[\frac{EI}{2} \left(\frac{d\theta}{dx} \right)^2 - P \cos \theta \right] = 0$$

The last expression is immediately integrable,

$$\frac{EI}{2} \left(\frac{d\theta}{dx} \right)^2 = P \cos \theta + c_1 \tag{6.30}$$

For a pinned column, θ is unknown, but $\theta'(x)$ is zero, at the ends of the column. If α is the unknown slope at the column ends, the vanishing of the moment there is satisfied by taking c_1 such that

$$\frac{EI}{2} \left(\frac{d\theta}{dx} \right)^2 = P \cos \theta - P \cos \alpha \tag{6.31}$$

Hence

$$\frac{\mathrm{d}\theta}{\mathrm{d}x} = \left(\frac{2P}{EI}\right)^{1/2}(\cos\theta - \cos\alpha)^{1/2}$$

$$= \left(\frac{4P}{EI}\right)^{1/2}\left(\sin^2\frac{\alpha}{2} - \sin^2\frac{\theta}{2}\right)^{1/2} \tag{6.32}$$

Then let us introduce a new variable ϕ, defined by

$$\sin\frac{\theta}{2} = \sin\frac{\alpha}{2}\sin\phi = q\sin\phi \tag{6.33}$$

This substitution enables us to find that

$$\frac{\mathrm{d}\phi}{\mathrm{d}x} = \left(\frac{4P}{EI}\right)^{1/2}(q\cos\phi)\frac{\mathrm{d}\phi}{\mathrm{d}\theta} \tag{6.34}$$

From the transformation (6.33) we can find

$$\frac{\mathrm{d}\theta}{\mathrm{d}\phi} = \frac{2q\cos\phi}{\sqrt{1 - q^2\sin^2\phi}} \tag{6.35}$$

Then, if we combine equations (6.34) and (6.35), we find a form,

$$\frac{\mathrm{d}\phi}{\mathrm{d}x} = \left(\frac{P}{EI}\right)^{1/2}(1 - q^2\sin^2\phi)^{1/2} \tag{6.36}$$

which can be implicitly integrated, to yield

$$\left(\frac{P}{EI}\right)^{1/2}\int_0^L \mathrm{d}x = \sqrt{\frac{PL^2}{EI}} = \int_{-\pi/2}^{\pi/2}\frac{\mathrm{d}\phi}{\sqrt{1 - q^2\sin^2\phi}}$$

or

$$\sqrt{\frac{P}{P_E}} = \frac{2}{\pi}\int_0^{\pi/2}\frac{\mathrm{d}\phi}{\sqrt{1 - q^2\sin^2\phi}} \tag{6.37}$$

The integral in equation (6.37) is an elliptic integral of the first kind, and its values are tabulated [5].

We can also compute the maximum deflection in terms of P/P_E. Note from equation (5.7) that $\sin \theta = dw/dx$, so that the differential equation (6.28) can be written as

$$EI \frac{d^2\theta}{dx^2} + P \frac{dw}{dx} = 0 \tag{6.38}$$

which in turn can be integrated,

$$w = -\frac{EI}{P} \frac{d\theta}{dx} + c_2 \tag{6.39}$$

At the column ends, both the deflection and moment vanish, so that $c_2 = 0$. From equation (6.32) we can solve for $d\theta/dx$, i.e.,

$$w = -\left(\frac{2EI}{P}\right)^{1/2} (\cos \theta - \cos \alpha)^{1/2} \tag{6.40}$$

The maximum deflection occurs at $\theta = 0$, so that

$$w_{max} = -\left(\frac{2EI}{P}\right)^{1/2} (1 - \cos \alpha)^{1/2} \tag{6.41}$$

which can easily be brought into the form

$$\left(\frac{w_{max}}{L}\right)^2 = \frac{2}{\pi^2} \left(\frac{P_E}{P}\right) (1 - \cos \alpha)$$

$$= \left(\frac{2}{\pi}\right)^2 \left(\frac{P_E}{P}\right) q^2 \tag{6.42}$$

Then we can plot the load-deflection curve, i.e., P/P_E against w_{max}/L, through appropriate combination of equations (6.37, 6.42), using a set of tables to evaluate the integral in equation (6.37). A graph will be given in the next section.

6.4   Asymptotic analysis—Initial postbuckling behavior

In this section we shall develop an asymptotic approximation for the early postbuckling behavior of the column. The approximation will not only be useful in its own right, and for comparison with the elastica result, but will serve as a demonstration of the Koiter (asymptotic) approach to initial postbuckling behavior (see Ref. [6] and the appropriate citations in Ref. [7]).

We begin by writing the potential energy (5.13) in dimensionless form, using equation (5.9) for the curvature and equation (5.10) for the shortening. If we interpret $w(x)$ and x as dimensionless, having divided the corresponding physical quantities by the column length L, we find

$$P^{(\lambda)}(w) = \left(\frac{2L}{EI}\right)\pi = \int_0^1 \frac{(w'')^2}{1 - (w')^2}\,dx + 2\lambda \int_0^1 \left[(1 - (w')^2)^{1/2} - 1\right]dx$$

(6.43)

where

$$\lambda = \frac{PL^2}{EI} = \frac{\pi^2 P}{P_E}$$

(6.44)

To make the energy functional (6.43) more tractable, we expand the denominator of the first integral and the radical of the second, to achieve a form

$$P^{(\lambda)}(w) = P_2^{(\lambda)} + P_4^{(\lambda)}$$

(6.45)

where $P_k^{(\lambda)}$ represents a functional of the kth order, in this case

$$P_2^{(\lambda)} = \int_0^1 \left[(w'')^2 - \lambda(w')^2\right]dx$$

(6.46a)

$$P_4^{(\lambda)} = \int_0^1 \left[(w'')^2(w')^2 - \tfrac{1}{4}\lambda(w')^4\right]dx$$

(6.46b)

Then the criterion of Trefftz, expressed as a necessary condition for stability, would be [8]

$$P_2^{(\lambda)} \geq 0$$

(6.47)

Koiter also showed that the stability of the postbuckling behavior is determined by the stability at the critical point, where bifurcation buckling takes place. A necessary condition for stability at the critical point is

$$P_3^{(\lambda)}(w_1) = 0, \quad P_4^{(\lambda)}(w_1) \geq 0$$

(6.48)

where w_1 is the buckling mode shape. The corresponding sufficiency conditions will be demonstrated by example.

Basically what we shall be about is a development of asymptotic expressions for the energy and the load, in the neighborhood of the critical load, in terms of a small parameter ε which is used as the amplitude of the buckling mode. Thus we take

$$w = \varepsilon w_1 + \varepsilon^2 w_2 + \varepsilon^3 w_3 + \ldots \tag{6.49}$$

and we expect the load parameter λ to be given by

$$\lambda = \lambda_{cr}(1 + a\varepsilon + b\varepsilon^2 + \ldots) \tag{6.50}$$

where the parameters $a, b \ldots$ remain to be determined, and where, we shall also see, the load-deflection relation (6.50) may be determined by

$$\frac{\mathrm{d}P^{(\lambda)}(w)}{\mathrm{d}\varepsilon} = 0 \tag{6.51}$$

Then the precise nature of the stability will be determined by the parameters $a, b \ldots$, and may be interpreted in terms of whether $\lambda/\lambda_{cr} \gtrless 1$, i.e., whether the postbuckling configuration carries more or less load than the critical load.

To examine the column behavior we proceed as follows. First, as we are interested in loads λ near λ_{cr}, we can write

$$P_k^{(\lambda)} = P_k^{(\lambda_{cr})} + (\lambda - \lambda_{cr})\, P_k^{'(\lambda_{cr})} \tag{6.52}$$

As we shall demonstrate by repeated example, equation (6.50) is sufficiently accurate as shown, implying that we need not keep terms of order ε^5 in $P^{(\lambda)}(w)$, after use of the expansions (6.49, 6.50, 6.52). Thus we will write that

$$\begin{aligned}
P^{(\lambda)}(w) &= P_2^{(\lambda_{cr})} + (\lambda - \lambda_{cr})P_2^{'(\lambda_{cr})} + P_4^{(\lambda_{cr})} \\
&= \int_0^1 [(w'')^2 - \lambda_{cr}(w')^2]\, \mathrm{d}x - (\lambda - \lambda_{cr})\int_0^1 (w')^2\, \mathrm{d}x \\
&\quad + \int_0^1 [(w'')^2(w')^2 - \tfrac{1}{4}\lambda_{cr}(w')^4]\, \mathrm{d}x
\end{aligned} \tag{6.53}$$

Now we substitute the expansion (6.49):

$$P^{(\lambda)}(w) = \varepsilon^2 \int_0^1 [(w_1'')^2 - \lambda_{cr}(w_1')^2]\,\mathrm{d}x - \varepsilon^2(\lambda - \lambda_{cr})\int_0^1 (w_1')^2\,\mathrm{d}x$$

$$+ \varepsilon^3 \left\{ 2\int_0^1 [w_1''w_2'' - \lambda_{cr}w_1'w_2']\,\mathrm{d}x - 2(\lambda - \lambda_{cr})\int_0^1 w_1'w_2'\,\mathrm{d}x \right\}$$

$$+ \varepsilon^4 \left\{ \int_0^1 [(w_2'')^2 + 2w_1''w_3'' - \lambda_{cr}(w_2')^2 - 2\lambda_{cr}w_1'w_3']\,\mathrm{d}x \right.$$

$$- (\lambda - \lambda_{cr})\int_0^1 [(w_2')^2 + 2w_1'w_3']\,\mathrm{d}x + \int_0^1 [(w_1'')^2(w_1')^2$$

$$\left. - \tfrac{1}{4}\lambda_{cr}(w_1')^4]\,\mathrm{d}x \right\} + \dots \tag{6.54}$$

Before proceeding further, we point out here that the terms $w_1, w_2, w_3, \dots$ are determined as the solutions of the perturbations of the exact differential equation, which is, in an integrated form,

$$\frac{\mathrm{d}}{\mathrm{d}x}\frac{w''}{\sqrt{1 - (w')^2}} + \lambda w' = 0$$

or, for supported columns,

$$\frac{w''}{\sqrt{1 - (w')^2}} + \lambda w = 0 \tag{6.55}$$

These perturbed equations are

$$w_1'' + \lambda_{cr}w_1 = 0 \tag{6.56a}$$

$$w_2'' + \lambda_{cr}w_2 = -a\lambda_{cr}w_1 \tag{6.56b}$$

etc. In this instance, the perturbed equations are simple, and they will have obvious consequences for the functional (6.54). In fact, for example, $w_2 = 0$.[1]

---

[1] This is based, of course, on recognizing that $a = 0$ for this problem.

We also point out that in general we would have the problem of obtaining higher order perturbations to get the requisite terms in the energy expansion, e.g., we need w_3 to calculate the term ε^4 in the expansion (6.54). To avoid this, we will introduce a (somewhat arbitrary) orthogonality condition, that is, we shall insist that w_1, the buckling mode, be orthogonal to the higher modes w_2, $w_3 \ldots$ in some fashion.

For the present problem, we choose the orthogonality condition as

$$\int_0^1 w_1' w_k' \, dx = 0, \quad k = 2, 3, \ldots \tag{6.57}$$

We also note from the fourth order buckling equation (6.56a, 6.2) that the critical load can also be given as the Rayleigh quotient

$$\lambda_{cr} = \int_0^1 (w_1'')^2 \, dx \bigg/ \int_0^1 (w_1')^2 \, dx \tag{6.58}$$

Further, equations (6.56a) and (6.57) imply that

$$\int_0^1 w_1'' w_k'' \, dx = 0, \quad k = 2, 3 \ldots \tag{6.59}$$

Then by using equations (6.56-6.59) we can reduce the energy functional (6.54) to the simpler form of

$$P^{(\lambda)}(w_1) = -\varepsilon^2(\lambda - \lambda_{cr}) \int_0^1 (w_1')^2 \, dx$$

$$+ \varepsilon^4 \int_0^1 [(w_1'')^2(w_1')^2 - \tfrac{1}{4}\lambda_{cr}(w_1')^4] \, dx \tag{6.60}$$

Let us define the parameter b as

$$b = \frac{2}{\lambda_{cr}} \frac{\int_0^1 [(w_1'')^2(w_1')^2 - \dfrac{\lambda_{cr}}{4}(w_1')^4] \, dx}{\int_0^1 (w_1')^2 \, dx} \tag{6.61}$$

Then the potential energy can be written as

$$P^{(\lambda)}(w_1) = \left[\int_0^1 (w_1')^2 \, dx \right] \{ -(\lambda - \lambda_{cr})\varepsilon^2 + \tfrac{1}{2}\lambda_{cr}b\varepsilon^4 \} \qquad (6.62)$$

From equation (6.62) we observe that at the bifurcation point, the critical point, $\lambda = \lambda_{cr}$, we have the energy directly proportional to b. Thus, if b is positive, the critical state is stable. If b is negative, the critical state is unstable. If b vanishes, additional (higher order) terms in the expansion are required to determine stability.

If we now apply the condition (6.51) to the energy variation (6.62) we can easily find that

$$\lambda/\lambda_{cr} = 1 + b\varepsilon^2 \qquad (6.63)$$

Thus another measure of stability is whether the structure can carry any additional load after bifurcation ($b > 0$, stable) or not ($b < 0$, unstable).

Finally, for the pinned column, the buckling mode shape, normalized so that $w_1(\tfrac{1}{2}) = 1$ for convenience, is

$$w_1 = \sin \pi x \qquad (6.64)$$

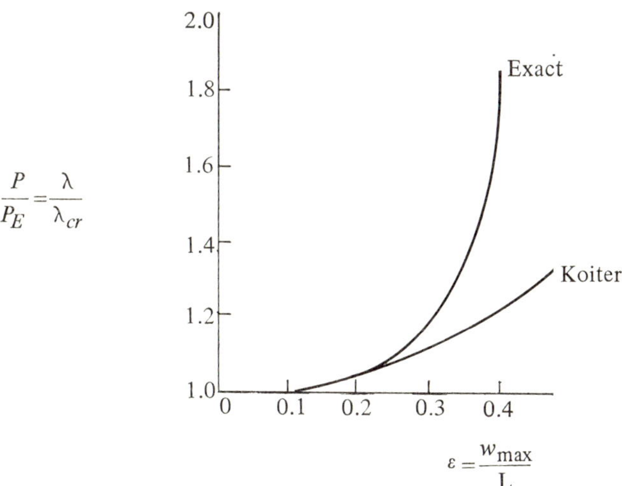

Fig. 2. Load-deflection curves for a column in the postbuckling state.

The bifurcation load is $\lambda_{cr} = \pi^2$, and the coefficient b may be calculated to show that

$$\lambda/\lambda_{cr} = 1 + (\pi^2/8)\varepsilon^2 \qquad (6.65)$$

In figure 2 we have plotted equation (6.65) as well as the exact result from the elastica analysis. Because of the normalization of w_1, with $w = \varepsilon w_1$, it is clear that ε in equation (6.65) plays the role of w_{max}/L. Note the good agreement. Also note from both analyses that the load is proportional to the square of the deflection. Thus it matters not whether the column bifurcates to the left or to the right, i.e., the sign of the buckling displacement is irrelevant. This is a feature of symmetric problems whose buckling is asymmetric.

6.5   Discussion of the Koiter asymptotic analysis

Considering that an elegant closed form solution is available, why bother, you may ask, with the ponderous asymptotic analysis? The answer has many parts. The first is that there are very few problems for which closed form solutions exist for the buckling and postbuckling problems. Thus it is useful to develop an approximate technique, and to check it against one of the known exact solutions.

While it is still true that the asymptotic analysis above was laborious, Budiansky and Hutchinson have developed an equivalent but neater formalism, based on the principle of virtual work [9, 10]. A side-by-side comparison of Koiter's original method and the Budiansky-Hutchinson approach was given for some arch problems by Dym [11]. While the final result was perforce the same for both analyses, the Budiansky-Hutchinson approach was shown to be much simpler to apply. In the next section we shall turn once more to the elastica to illustrate the other formalism with a now familiar result.

Before doing so we will briefly summarize the principal results of the asymptotic postbuckling analysis. They stem from the fact that after buckling at a bifurcation point, the load deflection behavior is reasonably well described by

$$\lambda/\lambda_{cr} = 1 + a\varepsilon + b\varepsilon^2 + \dots \qquad (6.66)$$

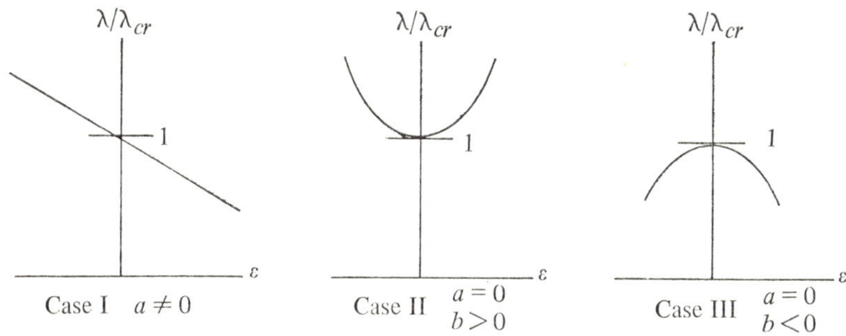

Fig. 3. Schematic postbuckling load-deflection curves for perfect structures.

The possible cases of interest are displayed in figure 3. In case I, the relevant terms in the expansion are $(1 + a\varepsilon)$, so that the postbuckling load-carrying capacity clearly is affected by the sign of the buckling displacement. For practical purposes, such a case would be classified as unstable. In the remaining two cases, the bifurcation is symmetric and the stability varies with the sign of b.

In addition to these interesting results, Koiter uncovered an even deeper set of results that have had a significant impact on the stability research of the last decade. He showed that if there was an imperfection in the shape of the structure, of very small amplitude $\bar{\varepsilon}$, and in the shape of the buckling mode, that equation (6.66) becomes

$$(1 - \lambda/\lambda_{cr})\varepsilon + a\varepsilon^2 + b\varepsilon^3 + \ldots = (\lambda/\lambda_{cr})\bar{\varepsilon} + \ldots \tag{6.67}$$

Thus, for a structure that is not geometrically perfect, we may obtain the load-deflection behavior in terms of the parameters of the perfect structure!

Further, it is then easy to show that in Case I (figure 3) on the 'side' where $a\varepsilon < 0$, that the imperfection causes a reduction in load-carrying capacity, below the bifurcation point. Then the maximum load is given by, for $a < 0$ say,

$$\lambda_m/\lambda_{cr} = 1 - 2(-a)^{1/2}\bar{\varepsilon}^{1/2} \tag{6.68}$$

A structure that exhibits this type of behavior is said to be *imperfection-sensitive*, as its usefulness is clearly degraded by imperfections.

In the second case, $a = 0$, $b > 0$, no deleterious effects are caused by the initial deviations, and the structure (in principle) does not experience a maximum load.

In the last case, $a = 0$, $b < 0$, the structure is again found sensitive to imperfections, here in the form

$$\lambda_m/\lambda_{cr} = 1 - 3 \left(\frac{-b}{4} \right)^{1/3} \bar{\varepsilon}^{2/3} \qquad (6.69)$$

Note the difference in exponents in equations (6.68, 6.69). The former is often said to refer to a 'quadratic' structure, the latter to a 'cubic,' because of the difference in exponents.

In closing this section we point out that the analysis leading to the asymptotic result (6.67) is valid for very small imperfections [6, 9]. While higher order terms in ε are kept, the products of $\varepsilon\bar{\varepsilon}$, $\bar{\varepsilon}^2$, etc., are generally accounted negligible.[1]

6.6 Asymptotic analysis—The Budiansky-Hutchinson approach

We proceed here by writing down the variation of the total potential energy. Thus, combining equations (6.45, 6.46) and carrying out the first variation, we find

$$\int_0^1 [w''(1 + (w')^2)\delta w'' + (w'')^2 w'\delta w' - \lambda(w' + \tfrac{1}{2}(w')^3)\delta w']dx = 0 \qquad (6.70)$$

Of course the first variation vanishes as it defines the equilibrium position. We now apply the displacement expansion (6.49) to the *unvaried* terms, i.e.,

$$\int_0^1 (w_1''\delta w'' - \lambda w_1' \, \delta w')dx + \varepsilon \int_0^1 (w_2''\delta w'' - \lambda w_2' \delta w')dx$$

---

[1] The qualifier 'generally' is inserted only for caution. If the Budiansky-Hutchinson formalism is used, no imperfection term other than $\lambda\bar{\varepsilon}/\lambda_{cr}$ appears in the (variational) virtual work statement. In Koiter's approach, the lowest order imperfection term in P^λ is likely to be $\lambda\varepsilon\bar{\varepsilon}/\lambda_{cr}$.

$$+ \, \varepsilon^2 \int_0^1 [(w_3'' + w_1''(w_1')^2)\delta w'' + w_1'(w_1'')^2 \delta w' - \lambda(w_3' + \tfrac{1}{2}(w_1')^3)\delta w'] \mathrm{d}x$$

$$+ \ldots = 0 \tag{6.71}$$

where we have divided through by ε once. From our anticipated result (6.66) we note that as $\varepsilon \to 0$, $\lambda \to \lambda_{cr}$, or

$$\int_0^1 (w_1''\delta w'' - \lambda_{cr}w_1'\delta w')\,\mathrm{d}x = 0 \tag{6.72}$$

Equation (6.72) is the variational statement of the eigenvalue problem that determines the critical load. Note also that the buckling displacement $w_1(x)$ would also be a legitimate substitution for $\delta w(x)$, so that equation (6.72) becomes

$$\int_0^1 [(w_1'')^2 - \lambda_{cr}(w_1')^2]\mathrm{d}x = 0 \tag{6.73}$$

which is simply the Rayleigh quotient. The same substitution would be equally valid in the variational statement (6.71), so that, taking note of the quotient (6.73), we have

$$(\lambda_{cr} - \lambda) \int_0^1 (w_1')^2\,\mathrm{d}x \; + \; \varepsilon \int_0^1 (w_2''w_1'' - \lambda w_2'w_1')\,\mathrm{d}x$$

$$+ \, \varepsilon^2 \int_0^1 [w_3''w_1'' - \lambda w_3'w_1' + 2(w_1'')^2(w_1')^2 - \lambda(\tfrac{1}{2})(w_1')^4]\,\mathrm{d}x + \ldots \tag{6.74}$$

We now choose as an orthogonality condition

$$\int_0^1 w_1'w_k'\,\mathrm{d}x = 0, \quad k = 2, 3 \ldots \tag{6.75}$$

and then it follows from equation (6.72) that, since the kth order mode shapes must also be kinematically admissible,

$$\int_0^1 w_1''w_k''\,\mathrm{d}x = 0 \tag{6.76}$$

If we apply the conditions (6.75, 6.76) to the variational statement (6.74) we find that

$$(\lambda_{cr} - \lambda) \int_0^1 (w_1')^2 \, dx + 2\varepsilon^2 \int_0^1 [(w_1'')^2(w_1')^2 - \tfrac{1}{4}\lambda(w_1')^4] \, dx + \ldots = 0 \tag{6.77}$$

Now in the term of order ε^2 it is perfectly permissible to replace λ with λ_{cr}, for $\lambda - \lambda_{cr} = 0(\varepsilon^2)$ is the result we expect to find. Then we do indeed obtain the result

$$\lambda/\lambda_{cr} = 1 + b\varepsilon^2 \tag{6.78}$$

where b is identically that constant found in the Koiter analysis, equation (6.61). Note we could also have found this result by direct substitution of the result (6.78) into both λ terms in equation (6.77).

To include imperfections in the above analysis, it is not difficult to show that the (dimensionless) curvature change and shortening are [1] [6]

$$\frac{1}{\rho} - \frac{1}{\rho_0} = \frac{w_0'' + w''}{\sqrt{1 - (w_0' + w')^2}} - w_0''$$

$$\frac{du}{dx} = (1 - (w_0' + w')^2)^{1/2} - 1 \tag{6.79}$$

Here $w(x)$ is the *additional deflection*, measured from the deformed (initially) column centerline. If we then modify appropriately the energy functional (6.43, 6.46) and carry out the first order variation, we get the following variational statement of the problem,

$$\int_0^1 [(1 + (w')^2 + 2w_0'w_1')w'' \, \delta w'' + \tfrac{1}{2}w_0''(w')^2 \, \delta w''$$

$$+ (w_0''w'' + (w'')^2)w' \, \delta w' + w_0'(w'') \, \delta w'$$

$$- \lambda w_0'(1 + \tfrac{3}{2}(w')^2) \, \delta w' - \lambda(w' + \tfrac{1}{2}(w')^3) \, \delta w'] \, dx = 0 \tag{6.80}$$

---

[1] One can simply use the curvature expression on p. 72 sequentially. Thus, initial state, $x_1^0 = x$, $x_3^0 = w_0(x)$ and $dl^0 = \sqrt{1 + (w_0')^2} \simeq dx$. For the deformed state, $x_1' = x + u$, $x_3' = w_0 + w$, and the usual incompressibility condition.

In developing equation (6.80) we have retained only first power terms of $w_0(x)$ by itself, and with products of $w(x)$. In addition to the expansion (6.49), we now define the imperfection amplitude by

$$w_0(x) = \bar{\varepsilon} w_1(x) \tag{6.81}$$

Substituting both (6.49) and (6.81) into the variational statement for the unvaried terms yields

$$\varepsilon \int_0^1 (w_1'' \, \delta w'' - \lambda w_1' \, \delta w') \, dx + \varepsilon^2 \int_0^1 (w_2'' \, \delta w'' - \lambda w_2' \, \delta w') \, dx$$

$$+ \, \varepsilon^3 \int_0^1 [w_3'' \, \delta w'' - \lambda w_3' \, \delta w' + w_1''(w_1')^2 \, \delta w'' + w_1'(w_1'')^2 \, \delta w'$$

$$- \tfrac{1}{2}\lambda(w_1')^3 \, \delta w'] \, dx - \lambda \bar{\varepsilon} \int_0^1 w_1' \, \delta w' \, dx + \ldots = 0 \tag{6.82}$$

As pointed out earlier, we are interested in very small imperfections. Thus we will assume that the following limits hold (see also our expected result, equation (6.67))

$$\lim_{\varepsilon \to 0} (\lim_{\bar{\varepsilon} \to 0} \lambda) = \lambda_{cr}, \quad \lim_{\varepsilon \to 0} \lambda = 0 \text{ if } \bar{\varepsilon} \neq 0. \tag{6.83}$$

If we then apply this first limit, we obtain the variational statement of the buckling problem (6.72) as before. We can then also replace $\delta w(x)$ by $w_1(x)$, invoke the orthogonality conditions (6.75, 6.76), and use the Rayleigh quotient (6.73) to find

$$\varepsilon(\lambda_{cr} - \lambda) \int_0^1 (w_1')^2 \, dx + \varepsilon^3 \int_0^1 [2(w_1'')^2(w_1')^2 - \tfrac{1}{2}\lambda(w_1')^4] \, dx$$

$$+ \ldots = \lambda \bar{\varepsilon} \int_0^1 (w_1')^2 \, dx \tag{6.84}$$

Then we continue as previously, replacing λ with λ_{cr} in the ε^3 term, and using the definition of b, to find, finally,

$$(1 - \lambda/\lambda_{cr})\varepsilon + b\varepsilon^3 = (\lambda/\lambda_{cr})\bar{\varepsilon} \tag{6.85}$$

This result can be compared to the Southwell-type result obtained from the linearized analysis, equation (6.22b).

6.7   Summary

We have devoted this chapter to a study of the buckling and postbuckling behavior of pinned-pinned columns. We have demonstrated a number of (linearized) buckling criteria and explored a number of approaches to examine the postbuckling behavior. The concept of geometric imperfections, and the possible degrading effects they can cause, has been introduced, and related to the stability of the bifurcation point.

Note that column buckling is a bifurcation phenomenon, except for an imperfect column. The maximum load of an imperfect column is a limit point buckling process. We shall see some limit point buckling of perfect structures when we discuss arches.

References

[1] H. Ziegler: *Principles of Structural Stability.* Blaisdell Publishing Co., Waltham, Mass., 1968.

[2] N. J. Hoff: *The Analysis of Structures.* John Wiley and Sons, 1956.

[3] R. M. Rivello: *Theory and Analysis of Flight Structures.* McGraw-Hill Book Company, New York, 1969.

[4] C. L. Dym and I. H. Shames: *Solid Mechanics: A Variational Approach.* McGraw-Hill Book Company, New York, 1973.

[5] M. Abramowitz and L. A. Stegun: *Handbook of Mathematical Functions.* Government Printing Office, Washington, D.C., 1965.

[6] W. T. Koiter: *The Stability of Elastic Equilibrium.* (A translation of a doctoral dissertation at Technische Hooge School at Delft, 1945), Technical Report AFFDL TR-70-25, Wright-Patterson Air Force Base, February 1970.

[7] J. W. Hutchinson and W. T. Koiter: Postbuckling Theory. *Applied Mechanics Reviews*, Vol. 23, No. 12, December 1970, p. 1353.

[8] W. T. Koiter: Elastic Stability and Postbuckling Behavior. *Proceedings of a Symposium on Nonlinear Problems*, University of Wisconsin Press, Madison, 1963.

[9] B. Budiansky: Postbuckling Behavior of Cylinders in Torsion. *Proceedings of the Second IUTAM Symposium on the Theory of Thin Shells*, edited by F. I. Niordson, Springer-Verlag, 1969.

[10] B. Budiansky and J. W. Hutchinson: Dynamic Buckling of Imperfection—Sensitive Structures. *Proceedings of the Eleventh International Congress of Applied Mechanics*, Julius Springer Verlag, 1964.

[11] C. L. Dym: Bifurcation Analyses for Shallow Arches. *Journal of the Engineering Mechanics Division*, ASCE, Vol. 99, No. EM2, April 1973.

7

Buckling and postbuckling of elastic plates

7.1 Energy criteria for rectangular plates

To establish the quadratic energy functional for the buckling of elastic plates, we shall simplify the energy functional (5.27) that we derived earlier. We shall consider here the buckling of rectangular plates, considering either singly or in combination compressive loads $\bar{N}_x$, $\bar{N}_y$ and a shear force $\bar{N}_{xy}$. In the reduction from the von Kármán theory to the linearized buckling equations, we shall make an assumption that parallels the classical theory of plates, i.e., we shall assume that the middle surface of the plate is *inextensible*. This implies, for example, that the first of equations (5.21) reduces to

$$\varepsilon_{11} = -z\,\frac{\partial^2 w}{\partial x^2} \tag{7.1}$$

with

$$\frac{\partial u}{\partial x} + \frac{1}{2}\left(\frac{\partial w}{\partial x}\right)^2 = 0 \tag{7.2}$$

Thus the strain energy of equation (5.27) becomes

$$U = \frac{D}{2}\int_0^a\int_0^b \left\{\left(\frac{\partial^2 w}{\partial x^2} + \frac{\partial^2 w}{\partial y^2}\right)^2 + 2(1-v)\left[\left(\frac{\partial^2 w}{\partial x\partial y}\right)^2\right.\right.$$
$$\left.\left. - \frac{\partial^2 w}{\partial x^2}\,\frac{\partial^2 w}{\partial y^2}\right]\right\}\,\mathrm{d}x\,\mathrm{d}y \tag{7.3}$$

where we have also used the moment-curvature relations (5.36, 5.37). It is also useful to note that the expression in the functional (7.3) that is multiplied by $2(1 - v)$ is called the Gaussian curvature [1]. The Gaussian curvature has the interesting property that, for a polygonal plate that is supported along all the edges ($w = 0$), and for a plate whose boundary is a smooth curve and whose edges are clamped, the contribution to the strain energy of this curvature term is identically zero. That is, for the conditions mentioned, it is possible to show by sequentially integrating-by-parts that

$$\int_R \int \left[\left(\frac{\partial^2 w}{\partial x \partial y} \right)^2 - \frac{\partial^2 w}{\partial x^2} \frac{\partial^2 w}{\partial y^2} \right] dx \, dy = 0 \tag{7.4}$$

For the potential of the edge loading (see equation (5.26)) we would have

$$V = \int_0^b \left[(\bar{N}_x u)_{x=a} - (\bar{N}_x u)_{x=0} + (\bar{N}_{xy} v)_{x=a} \right.$$

$$\left. - (\bar{N}_{xy} v)_{x=0} \right] dy + \int_0^a \left[(\bar{N}_y v)_{y=b} - (\bar{N}_y v)_{y=0} + (\bar{N}_{xy} u)_{y=b} \right.$$

$$\left. - (\bar{N}_{xy} u)_{y=0} \right] dx = \int_0^a \int_0^b \left[\bar{N}_x \frac{\partial u}{\partial x} + \bar{N}_{xy} \left(\frac{\partial u}{\partial y} + \frac{\partial v}{\partial x} \right) + \bar{N}_y \frac{\partial v}{\partial y} \right] dx \, dy$$

$$\tag{7.5}$$

In view of the inextensibility of the middle surface, equation (7.5) can be rewritten in the form

$$V = -\frac{1}{2} \int_0^a \int_0^b \left[\bar{N}_x \left(\frac{\partial w}{\partial x} \right)^2 + 2 \bar{N}_{xy} \frac{\partial w}{\partial x} \frac{\partial w}{\partial y} + \bar{N}_y \left(\frac{\partial w}{\partial y} \right)^2 \right] dx \, dy \tag{7.6}$$

Thus the combination of equations (7.3) and (7.6) is a quadratic energy functional that can be used to investigate the buckling of rectangular plates. We also point out that if we reduced the combination of these functionals to one-dimensional problems, we would essentially reproduce the column functional (6.1).

The vanishing of the first variation of the combined functional

$$\pi = \frac{D}{2}\int_0^a\int_0^b \left\{ (\nabla^2 w)^2 + 2(1-v)\left[\left(\frac{\partial^2 w}{\partial x \partial y}\right)^2 \right.\right.$$

$$\left.\left. - \frac{\partial^2 w}{\partial x^2}\frac{\partial^2 w}{\partial y^2}\right]\right\} dx\,dy - \frac{1}{2}\int_0^a\int_0^b \left[\bar{N}_x\left(\frac{\partial w}{\partial x}\right)^2 \right.$$

$$\left. + 2\bar{N}_{xy}\frac{\partial w}{\partial x}\frac{\partial w}{\partial y} + \bar{N}_y\left(\frac{\partial w}{\partial y}\right)^2 \right] dx\,dy \tag{7.7}$$

leads to the following eigenvalue problem defined by the differential equation

$$D\nabla^4 w + \bar{N}_x\frac{\partial^2 w}{\partial x^2} + 2\bar{N}_{xy}\frac{\partial^2 w}{\partial x \partial y} + \bar{N}_y\frac{\partial^2 w}{\partial y^2} = 0 \tag{7.8}$$

subject to the (possible) boundary conditions

$$\text{Either } M_v = 0 \text{ or } \frac{\partial w}{\partial v} \text{ is specified} \tag{7.9a}$$

and

$$\text{Either } Q_v + \frac{\partial M_{vs}}{\partial s} - \bar{N}_v\frac{\partial w}{\partial v} - \bar{N}_{vs}\frac{\partial w}{\partial s} = 0$$

$$\text{or } w \text{ is specified} \tag{7.9b}$$

We observe that we could also have obtained equation (7.8) from the von Kármán plate equations by linearizing the deflection and stress function about the pre-stress $(\bar{N}_x, \bar{N}_y, \bar{N}_{xy})$ state. Thus we would assume that

$$F = -\tfrac{1}{2}\bar{N}_x y^2 + \bar{N}_{xy}xy - \tfrac{1}{2}\bar{N}_y x^2 + \tilde{F}(x, y)$$

$$w = \tilde{w}(x, y) \tag{7.10}$$

Then, upon substituting equations (7.10) into equation (5.42) and neglecting squares and productions of the perturbed quantities $\tilde{w}$ and $\tilde{F}$, we would obtain equation (7.8).

In the same way that we solved the column eigenvalue problem, we will now consider the uni-axial compression of a simply supported rectangular plate. Thus we wish the eigenvalues of the equation

$$D\nabla^4 w + \overline{N}_x \frac{\partial^2 w}{\partial x^2} = 0 \tag{7.11}$$

subject to the boundary conditions

$$w = M_x = 0 \quad \text{at} \quad x = 0,a$$

$$w = M_y = 0 \quad \text{at} \quad y = 0,b$$

which can be put in the form

$$w = \frac{\partial^2 w}{\partial x^2} = 0 \quad \text{at} \quad x = 0,a$$

$$w = \frac{\partial^2 w}{\partial y^2} = 0 \quad \text{at} \quad y = 0,b \tag{7.12}$$

Motivated by the strong resemblance of the problem defined by equations (7.11, 7.12) to the column problem, we will attempt a solution in the form

$$w = W_{mn} \sin \frac{m\pi x}{a} \sin \frac{n\pi y}{b} \tag{7.13}$$

The solution (7.13) satisfies all the boundary conditions, and, upon substitution into the differential equation, yields

$$\left\{ D \left[\left(\frac{m\pi}{a} \right)^2 + \left(\frac{n\pi}{b} \right)^2 \right]^2 - \overline{N}_x \left(\frac{m\pi}{a} \right)^2 \right\} \times$$

$$\times W_{mn} \sin \frac{m\pi x}{a} \sin \frac{n\pi y}{b} = 0$$

Thus the requirement for a non-trivial solution is that

$$\overline{N}_x = D \left(\frac{a}{m\pi} \right)^2 \left[\left(\frac{m\pi}{a} \right)^2 + \left(\frac{n\pi}{b} \right)^2 \right]^2 \tag{7.14}$$

Since $\bar{N}_x = h\bar{\sigma}_x$, and since we can use the definition of D (equation (5.36)), we can write

$$\bar{\sigma}_x = \frac{\pi^2 E}{12(1 - v^2)} \left(\frac{h}{b}\right)^2 \left[\left(\frac{mb}{a}\right)^2 + 2n^2 + n^4 \left(\frac{a}{mb}\right)^2\right] \tag{7.15}$$

Here we have a doubly infinite set of values of $\bar{\sigma}_x$, corresponding to the infinite possible values of both the wave numbers, m and n. The situation is not as simple as the column result, where we could take the lowest value of the one (variable) wave number. We do notice, however, that if the *aspect ratio* a/b is fixed in equation (7.15), then the change in $\bar{\sigma}_x$ with n is monotone, and increasing with increasing values of n. Hence we could reasonably take $n=1$ to minimize $\bar{\sigma}_x$ at least with respect to this variable. Then we can also write that

$$\bar{\sigma}_x = \frac{E\pi^2}{12(1 - v^2)} \left(\frac{h}{b}\right)^2 k_{cr} \tag{7.16}$$

where we have introduced the *buckling coefficient* k_{cr}, which is here

$$k_{cr} = \left(m \frac{b}{a} + \frac{1}{m} \frac{a}{b}\right)^2 \tag{7.17}$$

We have plotted k_{cr} against the aspect ratio a/b for different values of m in figure 1. It is clear that the value of m giving the smallest buckling coefficient depends on the aspect ratio. Thus for $a/b = 1$ we see that $m = 1$, but, for $a/b = 2$, m must be 2. As a/b gets larger, notice that the critical buckling coefficient approaches the value 4. Establishing m and n for the lowest critical loads for a given ratio a/b thus establishes the buckling mode for the plate. For instance with $a/b = 2$ and $m = 2$ there must be a nodal line at $x = a/2$ (see figure 2) and we say that the plate has two buckles. The curved lines in the diagram may be thought of as contour lines of the buckled shape, the full lines indicating an upward deflection, the dashed lines indicating a downward deflection. For $m = 3$ there will be three such buckles, etc. Finally note that when a/b is an integer, then $m = a/b$ for achieving the minimum value of

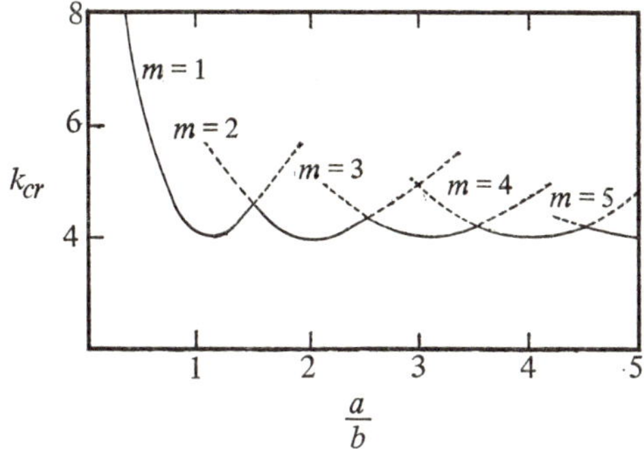

Fig. 1. Buckling coefficient for a simply-supported plate subject to uni-axial compression.

k_{cr}.[1] Thus for such cases the length of the plate (x direction) is divided into m buckles, i.e., m half-waves of length b.

Of course the above investigation has been of the equilibrium variety, rather than being of the energy type. We shall now briefly illustrate the energy approach, by considering the second variation of the functional (7.7), for

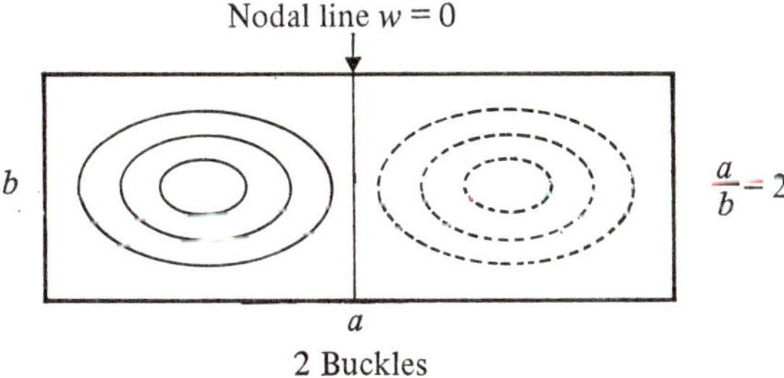

2 Buckles

Fig. 2. Buckling configuration for a simply-supported plate subject to uni-axial compression.

---

[1] We can reach the same conclusion by treating m as a continuous variable and we then minimize k_{cr} with respect to m. Thus:

$$\frac{\partial k_{cr}}{\partial m} = 0$$

This gives us $m = a/b$.

the case $\bar{N}_y = \bar{N}_{xy} = 0$. Since that functional is a quadratic functional it follows that the second variation (with η appearing for δw) is identical in form, that is

$$\delta^{(2)}\pi = \frac{D}{2}\int_0^a\int_0^b (\nabla^2\eta)^2\,dx\,dy - \tfrac{1}{2}\bar{N}_x\int_0^a\int_0^b \left(\frac{\partial\eta}{\partial x}\right)^2 dx\,dy \qquad (7.18)$$

In writing the result (7.18) we have dropped the Gaussian curvature contribution to the strain energy, as we are interested in a supported rectangular plate. Let us now assume that η (or δw) can be expressed as the doubly infinite sum

$$\eta = \sum_{m=1}^{\infty}\sum_{n=1}^{\infty} A_{mn}\sin\frac{m\pi x}{a}\sin\frac{n\pi y}{b} \qquad (7.19)$$

which leads immediately to the result that

$$\delta^{(2)}\pi = \frac{\pi^2}{8}\sum_{m=1}^{\infty}\sum_{n=1}^{\infty}\left\{\pi^2 Dab\left[\left(\frac{m}{a}\right)^2 + \left(\frac{n}{b}\right)^2\right]^2\right.$$

$$\left. - \frac{b\bar{N}_x}{a}\,m^2\right\}A_{mn}^2 \qquad (7.20)$$

It then easily follows that in order to insure the positivity of $\delta^{(2)}\pi$, that $\bar{N}_x$ must not exceed the right side of equation (7.14), so that we obtain the same result as previously.

One may also use the Trefftz criterion on the second variation given in equation (7.18). As with the column, the vanishing of $\delta^{(1)}[\delta^{(2)}\pi]$ leads to the eigenvalue problem that we have just solved, with w replaced by η, of course. Clearly we will then obtain the result (7.14) once more from that eigenvalue problem.

7.2 Buckling of circular plates

In this section we shall consider the buckling of circular plates. In order to do this we will first express the von Kármán plate equations in polar coordinates. Let us define radial and tangential displacements, u_r and u_θ respectively, by the transformations

$$u = u_r \cos \theta - u_\theta \sin \theta$$

$$v = u_r \sin \theta + u_\theta \cos \theta \tag{7.21}$$

We shall assume also that the transverse deflection is now $w = w(r, \theta)$. Then by proper use of the chain rule for differentiation,

$$\frac{\partial}{\partial x} = \cos \theta \, \frac{\partial}{\partial r} - \frac{1}{r} \sin \theta \, \frac{\partial}{\partial \theta}$$

$$\frac{\partial}{\partial y} = \sin \theta \, \frac{\partial}{\partial r} + \frac{1}{r} \cos \theta \, \frac{\partial}{\partial \theta} \tag{7.22}$$

and the strain-transformation relations of two-dimensional elasticity,

$$\varepsilon_{11} = \varepsilon_{rr} \cos^2 \theta + \varepsilon_{\theta\theta} \sin^2 \theta - 2\varepsilon_{r\theta} \sin \theta \cos \theta$$

$$\varepsilon_{22} = \varepsilon_{rr} \sin^2 \theta + \varepsilon_{\theta\theta} \cos^2 \theta + 2\varepsilon_{r\theta} \sin \theta \cos \theta$$

$$2\varepsilon_{12} = (\varepsilon_{rr} - \varepsilon_{\theta\theta}) \sin 2\theta + 2\varepsilon_{r\theta} \cos 2\theta \tag{7.23}$$

we would find

$$e_{rr} = \frac{\partial u_r}{\partial r} + \frac{1}{2} \left(\frac{\partial w}{\partial r} \right)^2 - z \frac{\partial^2 w}{\partial r^2}$$

$$\varepsilon_{\theta\theta} = \frac{1}{r} \frac{\partial u_\theta}{\partial \theta} + \frac{u_r}{r} + \frac{1}{2r^2} \left(\frac{\partial w}{\partial \theta} \right)^2 - z \left(\frac{1}{r} \frac{\partial w}{\partial r} + \frac{1}{r^2} \frac{\partial^2 w}{\partial \theta^2} \right)$$

$$2\varepsilon_{r\theta} = \frac{1}{r} \frac{\partial u_r}{\partial \theta} + \frac{\partial u_\theta}{\partial r} - \frac{u_\theta}{r} + \frac{1}{r} \frac{\partial w}{\partial r} \frac{\partial w}{\partial \theta}$$

$$- 2z \left(\frac{1}{r} \frac{\partial^2 w}{\partial r \partial \theta} - \frac{1}{r^2} \frac{\partial w}{\partial \theta} \right) \tag{7.24}$$

Equations (7.24) are thus the counterparts, in polar coordinates, of equations (5.21).

Then we can proceed exactly as we did in Section 5.2 when we derived the von Kármán theory for rectangular plates. Of principal interest to us here is the strain energy,

$$U = \frac{1}{2} \int_0^{2\pi} \int_0 \ (N_{rr}\varepsilon_{rr}^0 + N_{\theta\theta}\varepsilon_{\theta\theta}^0 + 2N_{r\theta}\varepsilon_{r\theta}^0$$

$$+ M_{rr}K_{rr} + M_{\theta\theta}K_{\theta\theta} + 2M_{r\theta}K_{r\theta})r\,dr\,d\theta \tag{7.25}$$

where we have used the notation

$$\varepsilon_{rr} = \varepsilon_{rr}^0 + zK_{rr} \tag{7.26}$$

etc., and the stress and moment resultants are defined as in the rectangular plate development (see equations (5.36) and (5.37)). The potential of the applied loads is, for $\bar{N}_{rr}$ in compression and $\bar{N}_{r\theta}$ in the negative θ direction,

$$V = \int_0^{2\pi} \bar{N}_{rr}(ru_r)_{r=b} \, d\theta + \int_0^{2\pi} \bar{N}_{r\theta}(ru_\theta)_{r=b} \, d\theta \tag{7.27}$$

We can also write the functional (7.27) as

$$V = \int_0^{2\pi} \int_0^b \left[\frac{\partial}{\partial r}(\bar{N}_{rr}ru_r) + \frac{\partial}{\partial r}(\bar{N}_{r\theta}ru_\theta) \right] dr\,d\theta$$

$$= \int_0^{2\pi} \int_0^b \left[\left(\frac{\partial \bar{N}_{rr}}{\partial r} + \frac{\bar{N}_{rr}}{r} \right) u_r + \bar{N}_{rr}\frac{\partial u_r}{\partial r} \right.$$

$$\left. + \left(\frac{\partial \bar{N}_{r\theta}}{\partial r} + \frac{\bar{N}_{r\theta}}{r} \right) u_\theta + \bar{N}_{r\theta}\frac{\partial u_\theta}{\partial r} \right] r\,dr\,d\theta \tag{7.28}$$

Now the applied loads must also satisfy the planar equations of elasticity, in polar coordinates. Hence, with $\bar{N}_{\theta\theta}$ also positive in compression,

$$\frac{\partial \bar{N}_{rr}}{\partial r} + \frac{\bar{N}_{rr}}{r} = \frac{\bar{N}_{\theta\theta}}{r} - \frac{1}{r}\frac{\partial \bar{N}_{r\theta}}{\partial \theta}$$

$$\frac{\partial \bar{N}_{r\theta}}{\partial r} + \frac{\bar{N}_{r\theta}}{r} = -\frac{1}{r}\frac{\partial \bar{N}_{\theta\theta}}{\partial \theta} - \frac{\bar{N}_{r\theta}}{r} \tag{7.29}$$

We also note that for a plate that is complete in the circumferential direction

$$-\int_0^{2\pi} \frac{\partial \overline{N}_{r\theta}}{\partial \theta}\, u_r\, \mathrm{d}\theta = \int_0^{2\pi} \overline{N}_{r\theta}\, \frac{\partial u_r}{\partial \theta}\, \mathrm{d}\theta$$

$$-\int_0^{2\pi} \frac{\partial \overline{N}_{\theta\theta}}{\partial \theta}\, u_\theta\, \mathrm{d}\theta = \int_0^{2\pi} \overline{N}_{\theta\theta}\, \frac{\partial u_\theta}{\partial \theta}\, \mathrm{d}\theta \tag{7.30}$$

Now we can combine equations (7.28, 7.29, 7.30) to find for the load potential the result

$$V = \int_0^{2\pi}\int_0^b \left[\overline{N}_{rr}\, \frac{\partial u_r}{\partial r} + \overline{N}_{\theta\theta}\left(\frac{1}{r}\, \frac{\partial u_\theta}{\partial \theta} + \frac{u_r}{r} \right) \right.$$

$$\left. + \overline{N}_{r\theta}\left(\frac{\partial u_\theta}{\partial r} - \frac{u_\theta}{r} + \frac{1}{r}\, \frac{\partial u_r}{\partial \theta} \right) \right] r\, \mathrm{d}r\, \mathrm{d}\theta \tag{7.31}$$

The form (7.31) is very useful to us, for we can now use the strain-displacement relations of the middle surface (e.g., equations (7.24) and (7.26)) to write the potential of the applied loads as

$$V = \int_0^{2\pi}\int_0^b \left[\overline{N}_{rr}\left(\varepsilon_{rr}^0 - \frac{1}{2}\left(\frac{\partial w}{\partial r}\right)^2 \right) \right.$$

$$\left. + \overline{N}_{\theta\theta}\left(\varepsilon_{\theta\theta}^0 - \frac{1}{2r^2}\left(\frac{\partial w}{\partial \theta}\right)^2 \right) + \overline{N}_{r\theta}\left(2\varepsilon_{r\theta}^0 - \frac{1}{r}\, \frac{\partial w}{\partial r}\, \frac{\partial w}{\partial \theta} \right) \right] r\, \mathrm{d}r\, \mathrm{d}\theta \tag{7.32}$$

Now, to consider only the buckling problem, we again make the assumption of inextensional behavior for the plate middle surface. Thus, with $\varepsilon_{rr}^0 = \varepsilon_{\theta\theta}^0 = \varepsilon_{r\theta}^0 = 0$ and using the moment-curvature relations for the circular plates, we can find that

$$\pi = \frac{D}{2}\int_0^{2\pi}\int_0^b \left[(\nabla^2 w)^2 + 2(1-\nu)\, \times \right.$$

$$\left. \left(\left(\frac{1}{r}\, \frac{\partial^2 w}{\partial r \partial \theta} - \frac{1}{r^2}\, \frac{\partial w}{\partial \theta} \right)^2 - \left(\frac{\partial^2 w}{\partial r^2} \right)\left(\frac{1}{r}\, \frac{\partial w}{\partial r} + \frac{1}{r^2}\, \frac{\partial^2 w}{\partial \theta^2} \right) \right) \right] r\, \mathrm{d}r\, \mathrm{d}\theta$$

$$-\frac{1}{2}\int_0^{2\pi}\int_0^b \left[\overline{N}_{rr}\left(\frac{\partial w}{\partial r}\right)^2 + \overline{N}_{\theta\theta}\left(\frac{1}{r}\, \frac{\partial w}{\partial \theta}\right)^2 + 2\overline{N}_{r\theta}\frac{1}{r}\, \frac{\partial w}{\partial \theta}\, \frac{\partial w}{\partial r} \right] r\, \mathrm{d}r\, \mathrm{d}\theta \tag{7.33}$$

Note the strong resemblance between the functional (7.33) and equation (7.7).

We will now consider the axisymmetric buckling of circular plates, under an applied load of value $\overline{N}_{rr}$. Then the functional (7.33) simplifies to

$$\pi = D\pi \int_0^b \left[\left(\frac{1}{r} \frac{dw}{dr} + \frac{d^2 w}{dr^2} \right)^2 - 2(1 - v) \frac{d^2 w}{dr^2} \left(\frac{1}{r} \frac{dw}{dr} \right) \right.$$
$$\left. - \alpha^2 \left(\frac{dw}{dr} \right)^2 \right] r \, dr \tag{7.34}$$

where

$$\alpha^2 = \overline{N}_{rr}/D \tag{7.35}$$

We shall use the quadratic functional (7.34) to illustrate the Trefftz criterion. Thus, because the energy is a quadratic functional, its second variation can be written as (with $\eta = \delta w$)

$$\delta^{(2)}\pi = D\pi \int_0^b \left[\left(\frac{d^2\eta}{dr^2} \right)^2 + \frac{1}{r^2} \left(\frac{d\eta}{dr} \right)^2 + 2v \frac{1}{r} \frac{d\eta}{dr} \frac{d^2\eta}{dr^2} \right.$$
$$\left. - \alpha^2 \left(\frac{d\eta}{dr} \right)^2 \right] r \, dr \tag{7.36}$$

The second variation ceases to be positive definite, according to the Trefftz criterion, when its first variation vanishes. Thus we obtain

$$\delta^{(1)}\left[\delta^{(2)}\pi\right] = 0 = \int_0^b \left[\frac{d^4\eta}{dr^4} + \frac{2}{r} \frac{d^3\eta}{dr^3} + \right.$$
$$\left. + \left(\alpha^2 - \frac{1}{r^2} \right) \frac{d^2\eta}{dr^2} + \left(\alpha^2 + \frac{1}{r^2} \right) \frac{1}{r} \frac{d\eta}{dr} \right] \delta\eta \, r \, dr$$

$$+ \text{ boundary terms} \tag{7.37}$$

Thus the governing differential equation is

$$r\,\eta^{IV} + 2\eta''' + \left(\alpha^2 r - \frac{1}{r} \right)\eta'' + \left(\alpha^2 + \frac{1}{r^2} \right)\eta' = 0 \tag{7.38}$$

where the primes denote ordinary differentiation with respect to r. We shall consider here the clamped plate, whose boundary conditions are

$$\eta = \eta' = 0 \quad \text{at } r = b \tag{7.39}$$

and the simply-supported plate, for which

$$r\eta'' + \nu\eta' = 0, \quad \eta = 0 \quad \text{at } r = b \tag{7.40}$$

Now the differential equation (7.38) can be immediately integrated once to yield

$$r\eta''' + \eta'' + \left(\alpha^2 r - \frac{1}{r}\right)\eta' = 0 \tag{7.41}$$

which is a Bessel equation for the dependent variable $\eta'(r)$ [2]. Thus we have started with a fourth order system (equation (7.38)), and have reduced it to a second order system for η'. Further, we have mentioned only two boundary conditions, i.e., either the pair (7.39) or the pair (7.40). This is because we are implicitly assuming that we are concerned with complete plates, and not with annuli. Then, because of the nature of the solutions to equation (7.41), we shall have to enforce regularity conditions at the origin. We mean by this that η and η' must be finite as $r \to 0$.

The regular solution of equation (7.41) is [2, 3]

$$\eta'(r) = c_1 J_1(\alpha r) \tag{7.42}$$

which is a Bessel function of order one, which goes to zero as $r \to 0$. Then

$$\eta(r) = c_2 - \frac{1}{\alpha} c_1 J_0(\alpha r) \tag{7.43}$$

where $J_0(\alpha r)$ is a Bessel function of order zero with the property $J_0(0) = 1$. Thus it is clear that our solution will be regular.

For the clamped plate we must satisfy equations (7.39). Thus we obtain

$$c_2 - \frac{1}{\alpha} c_1 J_0(\alpha b) = 0$$

$$c_1 J_1(\alpha b) = 0 \tag{7.44}$$

The first of these is easily satisfied, but the second requires — for a non-trivial solution to both equations — that

$$J_1(\alpha b) = 0 \tag{7.45}$$

The functions J_i are (essentially) periodic functions with almost uniformly spaced zeroes, that is, $J_1(\alpha b)$ does vanish whenever the argument takes on one of the values [3]

$$\alpha b = 0, 3.832, 7.016, 10.173, 13.323 \ldots \tag{7.46}$$

For our purposes, since $\alpha b = 0$ corresponds to a vanishing critical load, we may say that for the clamped plate a solution to the eigenvalue problem (7.41, 7.39) exists whenever

$$\alpha b = 3.832$$

or

$$\bar{N}_{rr} = 14.68 \, \frac{D}{b^2} \tag{7.47}$$

Thus for radial forces greater than the above value, the second variation (7.36) ceases to be positive definite, and a stable equilibrium configuration is not possible.

For the simply-supported plate, since

$$\frac{\mathrm{d}J_1(x)}{\mathrm{d}x} = J_0(x) - \frac{1}{x} \, J_1(x) \tag{7.48}$$

application of the zero-moment condition (7.40) yields the requirement that

$$(\alpha b)J_0(\alpha b) = (1 - v)J_1(\alpha b) \tag{7.49}$$

The roots of the transcendental equation (7.49) can be found by numerical analysis. For $v = 0.30$, the lowest root is

126

$$\alpha b = 2.049$$

or

$$\overline{N}_{rr} = 4.20 \frac{D}{b^2} \tag{7.50}$$

7.3   Postbuckling behavior of circular plates

In this section we shall consider the postbuckling behavior (axisymmetric) of the radially compressed state. The first variation of the potential energy can be written as

$$\delta^{(1)}\pi = \int_0^{2\pi} \int_0^b (N_{rr}\delta\varepsilon_{rr}^0 + N_{\theta\theta}\delta\varepsilon_{\theta\theta}^0$$

$$+ M_{rr}\delta K_{rr} + M_{\theta\theta}\delta K_{\theta\theta} + \overline{N}_{rr} \frac{d\delta u_r}{dr} + \frac{\overline{N}_{\theta\theta}}{r} \delta u_r) \, r\,dr\,d\theta = 0 \tag{7.51}$$

This is actually the same as a virtual work statement of the problem, and will be the starting point for our asymptotic analysis à la Budiansky and Hutchinson (see [4]).

We will adopt the following notation to simplify our analysis:

$$\varepsilon_{rr}^0 = e_{rr} + \frac{1}{2}\chi^2, \quad e_{rr} = \frac{du_r}{dr}, \quad \chi = \frac{dw}{dr}$$

$$\varepsilon_{\theta\theta}^0 = e_{\theta\theta}, \quad e_{\theta\theta} = \frac{u_r}{r} \tag{7.52}$$

Then the virtual work statement (7.51) can be written as

$$0 = \int_0^b (N_{rr}\delta e_{rr} + N_{\theta\theta}\delta e_{\theta\theta} + N_{rr}\chi\,\delta\chi + M_{rr}\delta K_{rr} + M_{\theta\theta}\delta K_{\theta\theta} + \overline{N}_{rr}\delta e_{rr}$$

$$+ \overline{N}_{\theta\theta}\delta e_{\theta\theta})r\,dr \tag{7.53}$$

Let us now introduce the following expansions, i.e.,

$$
\begin{bmatrix} u_r \\ w \\ \chi \\ N_{ij} \\ \\ M_{ij} \\ e_{ij} \\ K_{ij} \end{bmatrix} = \lambda \begin{bmatrix} u_0 \\ w_0 \\ \chi_0 \\ N_{ij}^0 \\ \\ M_{ij}^0 \\ e_{ij}^0 \\ K_{ij}^0 \end{bmatrix} + \varepsilon \begin{bmatrix} u_1 \\ w_1 \\ \chi_1 \\ N_{ij}^{(1)} \\ \\ M_{ij}^{(1)} \\ e_{ij}^{(1)} \\ K_{ij}^{(1)} \end{bmatrix} + \varepsilon^2 \begin{bmatrix} u_2 \\ w_2 \\ \chi_2 \\ N_{ij}^{(2)} \\ \\ M_{ij}^{(2)} \\ e_{ij}^{(2)} \\ K_{ij}^{(2)} \end{bmatrix} + \ldots \tag{7.54}
$$

where the first column on the right defines the prebuckling problem, the second the buckling problem, and so on. Remember that $\varepsilon \to 0$ implies $\lambda \to \lambda_{cr}$. If we then substitute for the unvaried quantities and order the resulting variational statement, we find

$$
0 = \int_0^b [(\lambda N_{rr}^0 + \overline{N}_{rr})\delta\varepsilon_{rr} + (\lambda N_{\theta\theta}^0 + \overline{N}_{\theta\theta})\delta e_{\theta\theta} + \lambda^2 N_{rr}^0 \overline{N}_{rr} \chi_0 \delta\chi
$$

$$
+ \lambda M_{rr}^0 \delta K_{rr} + \lambda M_{\theta\theta}^0 \delta K_{\theta\theta}] \, r \, dr + \varepsilon \int_0^b [N_{rr}^{(1)} \delta e_{rr}
$$

$$
+ N_{\theta\theta}^{(1)} \delta e_{\theta\theta} + \lambda(N_{rr}^0 \chi_1 + N_{rr}^{(1)} \chi_0)\delta\chi + M_{rr}^{(1)} \delta K_{rr} + M_{\theta\theta}^{(1)} \delta K_{\theta\theta}] \, r \, dr
$$

$$
+ \varepsilon^2 \int_0^b [N_{rr}^{(2)} \delta e_{rr} + N_{\theta\theta}^{(2)} \delta e_{\theta\theta} + (\lambda N_{rr}^0 \chi_2 + N_{rr}^{(1)} \chi_1 + \lambda N_{rr}^{(2)} \chi_0)\delta\chi
$$

$$
+ M_{rr}^{(2)} \delta K_{rr} + M_{\theta\theta}^{(2)} \delta K_{\theta\theta}] r \, dr + \varepsilon^3 \int_0^b [N_{rr}^{(3)} \delta e_{rr} + N_{\theta\theta}^{(3)} \delta e_{\theta\theta}
$$

$$
+ (\lambda N_{rr}^0 \chi_3 + N_{rr}^{(1)} \chi_2 + N_{rr}^{(2)} \chi_1 + \lambda N_{rr}^{(3)} \chi_0)\delta\chi + M_{rr}^{(3)} \delta K_{rr}
$$

$$
+ M_{\theta\theta}^{(3)} \delta K_{\theta\theta}] r \, dr + \ldots \tag{7.55}
$$

The first grouping of terms above defines the *prebuckling* problem. We shall insist that the prebuckling state be (for simplicity) a *linear membrane* state. The first statement requires that we drop the product term $\lambda^2 N_{rr}^0 \chi_0$, or, equivalently, we set $\chi_0 = 0$. The second condition requires $M_{rr}^0 = M_{\theta\theta}^0 = 0$, which thus also requires that $K_{rr}^0 = K_{\theta\theta}^0 = 0$. Then by setting

$$
\int_0^b [(\lambda N_{rr}^0 + \overline{N}_{rr})\delta e_{rr} + (\lambda N_{\theta\theta}^0 + \overline{N}_{\theta\theta})\delta e_{\theta\theta}] r \, dr = 0 \tag{7.56}
$$

we obtain the prebuckling problem, whose solution is given as

$$\lambda N_{rr}^0 = - \overline{N}_{rr}, \quad \lambda N_{\theta\theta}^0 = - \overline{N}_{\theta\theta} = - \overline{N}_{rr}$$

$$u_0 = - \frac{\overline{N}_{rr}(1-v)}{\lambda Eh} r, \quad w_0 = 0 \tag{7.57}$$

With the vanishing of the prebuckling statement, and with $\chi_0 = 0$, we will have left the following expansion:

$$\int_0^b (N^{(1)}\delta e + M^{(1)}\delta K + \lambda N_{rr}^0 \chi_1 \delta\chi) r\,dr + \varepsilon \int_0^b [N^{(2)}\delta e + M^{(2)}\delta K$$

$$+ (\lambda N_{rr}^0 \chi_2 + N_{rr}^{(1)}\chi_1)\delta\chi]r\,dr + \varepsilon^2 \int_0^b [N^{(3)}\delta e + M^{(3)}\delta K + (\lambda N_{rr}^0 \chi_3$$

$$+ N_{rr}^{(1)}\chi_2 + N_{rr}^{(2)}\chi_1)\delta\chi]r\,dr + \ldots = 0 \tag{7.58}$$

Note that in equation (7.58) we have introduced the symbolic notation

$$N^{(i)}\delta e = N_{rr}^{(i)}\delta e_{rr} + N_{\theta\theta}^{(i)}\delta e_{\theta\theta}$$

$$M^{(i)}\delta K = M_{rr}^{(i)}\delta K_{rr} + M_{\theta\theta}^{(i)}\delta K_{\theta\theta} \tag{7.59}$$

Now the buckling problem may be evaluated in one of two ways. If the variation of the zeroth order term in the expansion (7.58) is set to zero, since $\lambda \to \lambda_{cr}$ as $\varepsilon \to 0$, it follows that the buckling problem is the set of Euler-Lagrange equations of

$$\int_0^b (N^{(1)}\delta e + M^{(1)}\delta K + \lambda_{cr} N_{rr}^0 \chi_1 \delta\chi) r\,dr = 0 \tag{7.60}$$

Alternatively, we may define the critical load by the Rayleigh quotient obtained by letting $\delta e \to e_1$, $\delta K \to K_1$ and $\delta\chi \to \chi_1$ in equation (7.60):

$$\lambda_{cr} = - \frac{\displaystyle\int_0^b (N^{(1)}e^{(1)} + M^{(1)}K^{(1)})r\,dr}{\displaystyle N_{rr}^0 \int_0^b \chi_1^2 r\,dr} \tag{7.61}$$

129

In all the variational statements above, we will now replace the kinematic variations $(\delta e, \delta K, \delta \chi)$ by their buckling counterparts. Also, we will choose the orthogonality condition for this problem as

$$\int_0^b (N^{(1)}e^{(k)} + M^{(1)}K^{(k)})r\,dr = 0, \quad k = 2, 3 \ldots \tag{7.62}$$

As a consequence of equations (7.62) and (7.60), it must also be true that

$$\int_0^b \chi_1 \chi_m r\,dr = 0 \tag{7.63}$$

Before applying equation (7.62) we must also note that since

$$N_{rr} = C(e_{rr} + \frac{1}{2}\chi^2 + ve_{\theta\theta})$$

$$N_{\theta\theta} = C\left(ve_{rr} + \frac{v}{2}\chi^2 + e_{\theta\theta}\right) \tag{7.64}$$

it follows that some of the terms in equation (7.58) need to be recast before that orthogonality condition can be easily applied. In particular

$$N^{(2)}\delta e \to N_{rr}^{(2)}e_{rr}^{(1)} + N_{\theta\theta}^{(2)}e_{\theta\theta}^{(1)} = N_{rr}^{(1)}e_{rr}^{(2)} + N_{\theta\theta}^{(1)}e_{\theta\theta}^{(2)} + \tfrac{1}{2}N_{rr}^{(1)}\chi_1^2 \tag{7.65a}$$

$$N^{(3)}\delta e \to N_{rr}^{(3)}e_{rr}^{(1)} + N_{\theta\theta}^{(3)}e_{\theta\theta}^{(1)} = N_{rr}^{(1)}e_{rr}^{(3)} + N_{\theta\theta}^{(1)}e_{\theta\theta}^{(3)} + N_{rr}^{(1)}\chi_1\chi_2 \tag{7.65b}$$

Since the moment curvature relations are linear, a simpler result holds, i.e.,

$$M^{(k)}\delta K \to M^{(k)}K^{(1)} = M^{(1)}K^{(k)} \tag{7.66}$$

Now, in the expansion (7.58), we can replace the variations, use the Rayleigh quotient, and apply the orthogonality conditions. The resulting expansion is

$$(\lambda - \lambda_{cr})N_{rr}^0\int_0^b \chi_1^2\,r\,dr + \varepsilon\int_0^b \frac{3}{2}N_{rr}^{(1)}\chi_1^2\,r\,dr$$

$$+ \varepsilon^2\int_0^b (2N_{rr}^{(1)}\chi_1\chi_2 + N_{rr}^{(2)}\chi_1^2)r\,dr + \ldots = 0 \tag{7.67}$$

It then appears that the expansion (7.67) can be written as

$$\frac{\lambda}{\lambda_{cr}} = 1 + a\varepsilon + b\varepsilon^2 + \ldots \tag{7.68}$$

where

$$a = -\frac{3}{2\lambda_{cr}} \frac{\displaystyle\int_0^b N_{rr}^{(1)}\chi_1^2\, r\, dr}{\displaystyle\int_0^b N_{rr}^0 \chi_1^2\, r\, dr} \tag{7.69a}$$

$$b = -\frac{1}{\lambda_{cr}} \frac{\displaystyle\int_0^b [2N_{rr}^{(1)}\chi_1\chi_2 + N_{rr}^{(2)}\chi_1^2]\, r\, dr}{\displaystyle\int_0^b N_{rr}^0 \chi_1^2\, r\, dr} \tag{7.69b}$$

The equations that govern the first and second order expressions in the coefficients above may be obtained by setting to zero the appropriate variations of equation (7.58), or by perturbing the exact equations of equilibrium obtainable from the variational equation (7.51). In either case those equations are

$$0(\varepsilon) \qquad \frac{d}{dr}(rN_{rr}^{(1)}) - N_{\theta\theta}^{(1)} = 0 \tag{7.70a}$$

$$\frac{d}{dr}\left[\frac{d}{dr}(rM_{rr}^{(1)}) - M_{\theta\theta}^{(1)} + \lambda_{cr}N_{rr}^0(r\chi_1)\right] = 0 \tag{7.70b}$$

$$0(\varepsilon^2) \qquad \frac{d}{dr}(rN_{rr}^{(2)}) - N_{\theta\theta}^{(2)} = 0 \tag{7.71a}$$

$$\frac{d}{dr}\left[\frac{d}{dr}(rM_{rr}^{(2)}) - M_{\theta\theta}^{(2)} + \lambda_{cr}N_{rr}^0(r\chi_2) + N_{rr}^{(1)}(r\chi_1)\right] = 0 \tag{7.71b}$$

Although there is a strong structural similarity between these pairs of equations, they are not identical. This would be very evident if the equations were written in terms of displacements.

We shall attempt a solution for the clamped plate. Note that in addition to the out-of-plane boundary conditions (see equations (7.39))

$$w_i' = 0, \quad w_i = 0 \quad \text{at} \quad r = b \tag{7.72}$$

we will also have an in-plane condition, here to be

$$N_{rr}^{(i)} = 0 \quad \text{at} \quad r = b \tag{7.73}$$

For the buckling problem, we note that

$$N_{rr}^{(1)} = C\left(u_1' + \frac{v}{r}\,u_1\right), \quad N_{\theta\theta}^{(1)} = C\left(vu_1' + \frac{1}{r}\,u_1\right)$$

$$\chi_1 = w_1', \quad M_{rr}^{(1)} = -D\left(w_1'' + \frac{v}{r}\,w_1'\right)$$

$$M_{\theta\theta}^{(1)} = -D\left(vw_1'' + \frac{1}{r}\,w_1'\right) \tag{7.74}$$

where the primes are used once more for the ordinary derivatives with respect to the radial coordinate. Then the buckling problem differential equations reduce to

$$r\,u_1'' + u_1' - \frac{1}{r}\,u_1 = 0 \tag{7.75a}$$

$$r\,w_1''' + w_1'' - \left(\frac{\lambda_{cr}\,N_{rr}^0}{D}\,r + \frac{1}{r}\right)w_1' = 0 \tag{7.75b}$$

The second of these is our familiar buckling equation, for which we have already obtained the solution (equations (7.42, 7.43, 7.47)),

$$w_1(r) = \frac{c_1}{\alpha}\left[J_0(\alpha b) - J_0(\alpha r)\right] \tag{7.76a}$$

$$-\lambda_{cr}N_{rr}^0 = 14.68\,\frac{D}{b^2} = \alpha^2 D \tag{7.76b}$$

The constant c_1 is arbitrary, as befits the amplitude of the buckled shape, for an eigenvalue problem. We will normalize it so that the buckling deflection is equal to the plate thickness h at $r = 0$. Then we will have

$$w_1 = h\frac{J_0(\alpha b) - J_0(\alpha r)}{J_0(\alpha b) - 1} \tag{7.76c}$$

The membrane equation has the solution, for finite u_1 at $r = 0$, $u_1 = c_2 r$. However, when we enforce the boundary condition (7.73) for $N_{rr}^{(1)}$ we quickly find that $c_2 = 0$. Hence, during buckling,

$$u_1 = N_{rr}^{(1)} = 0. \tag{7.77}$$

It follows immediately, from equation (7.69a), that $a = 0$. This should not be surprising for the plate should not be sensitive to the sign of its buckling displacement. The result (7.77) also simplifies equation (7.69b) to the form

$$b = -\frac{1}{\lambda_{cr}N_{rr}^0}\frac{\displaystyle\int_0^b (N_{rr}^{(2)}\chi_1^2)r\,\mathrm{d}r}{\displaystyle\int_0^b \chi_1^2 r\,\mathrm{d}r}$$

or

$$b = -\frac{C}{\lambda_{cr}N_{rr}^0}\frac{\displaystyle\int_0^b [(e_{rr}^{(2)} + ve_{\theta\theta}^{(2)})\chi_1^2 + \tfrac{1}{2}\chi_1^4]r\,\mathrm{d}r}{\displaystyle\int_0^b \chi_1^2 r\,\mathrm{d}r} \tag{7.78}$$

Thus, to obtain the coefficient b, we need only evaluate u_2, since

$$e_{rr}^{(2)} = u_2', \quad e_{\theta\theta}^{(2)} = \frac{1}{r}u_2 \tag{7.79}$$

To obtain u_2, we transform equation (7.71a) by using equations (7.79) and the appropriate stress-strain equation. We find

$$r u_2'' + u_2' - \frac{1}{r}u_2 = -\frac{1-v}{2}\chi_1^2 - r\chi_1\chi_1' \tag{7.80}$$

where, remember, $\chi_1 = w_1'$. The appropriate boundary condition for $N_{rr}^{(2)}$ appears to be

$$u_2' + \frac{v}{r}\,u_2 + \frac{1}{2}\chi_1^2 = 0 \quad \text{at} \quad r = b$$

However, for the clamped plate, $w_1'(b) = 0$, so that the true boundary condition is

$$u_2' + \frac{v}{r}\,u_2 = 0 \quad \text{at} \quad r = b \tag{7.81}$$

The solution to equations (7.80) and (7.81) is

$$u_2 = \frac{\alpha h^2}{4[J_0(\alpha b) - 1]^2}\{(1 + v)J_0(\alpha r)J_1(\alpha r) - (1 - v)\alpha r\,J_0^2(\alpha b)$$

$$- v\alpha r\,[J_0^2(\alpha r) + J_1^2(\alpha r)]\} \tag{7.82}$$

Then we can also compute

$$N_{rr}^{(2)} = C\left(u_2' + \frac{1}{2}\,\chi_1^2 + \frac{v}{r}\,u_2\right)$$

$$= \frac{C\alpha^2 h^2(1 - v^2)}{4[J_0(\alpha b) - 1]^2}\left\{J_0^2(\alpha r) + J_1^2(\alpha r)\right.$$

$$\left. - \frac{1}{\alpha r}\,J_0(\alpha r)J_1(\alpha r) - J_0^2(\alpha b)\right\} \tag{7.83}$$

It is also easily verified that equation (7.83) satisfies the condition (7.81), as $J_1(\alpha b)$ is zero for the clamped plate. After some further manipulation the coefficient (7.78) can be put into the form

$$b = \frac{3(1 - v^2)}{[J_0(\alpha b) - 1]^2}\,\frac{I_1 - J_0^2(\alpha b)I}{I} \tag{7.84}$$

where

$$I = \int_0^{\alpha b} J_1^2(x)\,x\,\mathrm{d}x \tag{7.85a}$$

$$I_1 = \int_0^{\alpha b} [J_0^2(x) + J_1^2(x) - \frac{1}{x}\,J_0(x)J_1(x)]J_1^2(x)\,x\,\mathrm{d}x \tag{7.85b}$$

The first of these integrals is easily evaluated, and is [3]

$$I = \left[\tfrac{1}{2}x^2(J_0^2(x) + J_1^2(x)) - x J_0(x) J_1(x)\right]_0^{ab} = \tfrac{1}{2}(\alpha b)^2 J_0^2(\alpha b) \qquad (7.86)$$

The second integral, I_1, can be fiddled with to reduce it to the form

$$I_1 = \frac{1}{3}\int_0^{ab} \left(4x - \frac{1}{x}\right) J_1^4(x)\, dx \qquad (7.87)$$

which was integrated numerically, for $\alpha b = 3.832$, to yield $I_1 = 0.367$. Since there are tables [3] that give us the result $J_0(\alpha b) = -0.4028$, we can then calculate the coefficient b to yield the postbuckling load-deflection relation

$$\left(\frac{\lambda}{\lambda_{cr}}\right)_{\text{clamped}} = 1 + (0.2049)\varepsilon^2 \qquad (7.88)$$

A somewhat different approach to this problem has been given by Thompson and Lewis [5], although the final answers are virtually identical. They have also given the result for the simply-supported plate as

$$\left(\frac{\lambda}{\lambda_{cr}}\right)_{\text{pinned}} = 1 + (0.2702)\varepsilon^2 \qquad (7.89)$$

This result agrees very well with that obtained in a classic perturbation solution of Friedrichs and Stoker [6].

Before closing this section we also note that the actual buckling deflection is $w_{buck} = \varepsilon w_1$, and that we have normalized w_1 so that its maximum value (at $r = 0$) is equal to the plate thickness, h. Thus we can interpret the quantity ε in equations (7.88, 7.89) above as the maximum buckling deflection, rendered dimensionless with respect to the plate thickness.

7.4   Postbuckling of rectangular plates

In this section we shall consider a 'classical' analysis of the postbuckling behavior of rectangular plates. We will develop a Rayleigh-Ritz solution to the problem by approximating the complete (nonlinear) strain energy functional. Thus we will not be attempting an asymptotic analysis in this section.

From equation (5.27) we can reconstruct the strain energy for a square plate as

$$
U = \frac{Eh}{2(1 - v^2)} \int_0^a \int_0^a \left\{ \left[\frac{\partial u}{\partial x} + \frac{1}{2}\left(\frac{\partial w}{\partial x}\right)^2 \right]^2 + \left[\frac{\partial v}{\partial y} + \frac{1}{2}\left(\frac{\partial w}{\partial y}\right)^2 \right]^2 \right.
$$

$$
+ 2v \left[\frac{\partial u}{\partial x} + \frac{1}{2}\left(\frac{\partial w}{\partial x}\right)^2 \right]\left[\frac{\partial v}{\partial y} + \frac{1}{2}\left(\frac{\partial w}{\partial y}\right)^2 \right] + \frac{1 - v}{2}\left[\frac{\partial u}{\partial y} + \frac{\partial v}{\partial x} \right.
$$

$$
\left. + \frac{\partial w}{\partial x}\frac{\partial w}{\partial y} \right]^2 + \frac{h^2}{12}(\nabla^2 w)^2 \left. \right\} \, dx \, dy
\tag{7.90}
$$

In writing equation (7.90) we have noted that for the simply supported panel that we shall deal with here, there is no contribution to the bending energy from the Gaussian curvature term. Further, we shall consider here the behavior of the plate when *the edge displacement is prescribed*, rather than when the axial load is prescribed.

We will assume that the plate displacements can be represented as [7] [1]

$$
u = -\bar{u}x + u_{20}\sin\frac{2\pi x}{a} + u_{22}\sin\frac{2\pi x}{a}\cos\frac{2\pi y}{a}
$$

$$
v = \bar{v}y + v_{02}\sin\frac{2\pi y}{a} + v_{22}\cos\frac{2\pi x}{a}\sin\frac{2\pi y}{a}
$$

$$
w = w_{11}\sin\frac{\pi x}{a}\sin\frac{\pi y}{a}
\tag{7.91}
$$

The transverse component satisfies the simple support boundary conditions. You may verify that the in-plane displacements approximately satisfy the planar equations of equilibrium (equations (5.33a, b)). With the use of equations (7.91) the strain energy (7.90) becomes

---

1 This solution, adapted from an elasto-plastic solution given by Mayers and Budiansky [7], was given in the form presented here by Mayers [8]. See, also, Timoshenko and Gere [9], and Rivello [10].

$$
U = \frac{Ea^2 h}{2(1 - v^2)} \left\{ \bar{u}^2 - 2v\bar{u}\bar{v} + \bar{v}^2 + \frac{2\pi^2}{a^2} (u_{20}^2 + v_{02}^2) \right.
$$

$$
+ \frac{(3 - v)\pi^2}{a^2} (u_{22}^2 + v_{22}^2) + \frac{(1 + v)\pi^2}{a^2} u_{22} v_{22} +
$$

$$
+ \frac{1}{4} \left(\frac{w_{11}}{a} \right)^2 \left[(1 + v)\pi^2 (\bar{v} - \bar{u}) + \frac{(1 - v)\pi^3}{a} (u_{20} + v_{02}) \right.
$$

$$
\left. - \frac{\pi^3}{a} (u_{22} + v_{22}) + \frac{\pi^4}{3} \left(\frac{h}{a} \right)^2 \right] + \frac{5\pi^4}{64} \left(\frac{w_{11}}{a} \right)^4 \left. \right\} \tag{7.92}
$$

As $\bar{u}$ is prescribed, it plays a passive role in the energy extremization process. If we differentiate the energy equation (7.92) with respect to $\bar{v}$, u_{20}, v_{02}, u_{22}, v_{22}, and w_{11}, respectively, we obtain the following equations:

$$
\bar{v} = v\bar{u} - \frac{(1 + v)\pi^2}{8} \left(\frac{w_{11}}{a} \right)^2 \tag{7.93a}
$$

$$
\frac{u_{20}}{a} = -\frac{(1 - v)\pi}{16} \left(\frac{w_{11}}{a} \right)^2 \tag{7.93b}
$$

$$
\frac{v_{02}}{a} = -\frac{(1 - v)\pi}{16} \left(\frac{w_{11}}{a} \right)^2 \tag{7.93c}
$$

$$
(3 - v)\frac{u_{22}}{a} + (1 + v)\frac{v_{22}}{a} = \frac{\pi}{4} \left(\frac{w_{11}}{a} \right)^2 \tag{7.93d}
$$

$$
(1 + v)\frac{u_{22}}{a} + (3 - v)\frac{v_{22}}{a} = \frac{\pi}{4} \left(\frac{w_{11}}{a} \right)^2 \tag{7.93e}
$$

$$
\frac{5\pi^4}{8} \left(\frac{w_{11}}{a} \right)^3 + \left(\frac{w_{11}}{a} \right) \left[(1 + v)\pi^2 (\bar{v} - \bar{u}) + (1 - v)\pi^3 \left(\frac{u_{20}}{a} + \right. \right.
$$

$$
\left. \left. \frac{v_{02}}{a} \right) - \pi^3 \left(\frac{u_{22}}{a} + \frac{v_{22}}{a} \right) + \frac{\pi^4}{3} \left(\frac{h}{a} \right)^2 \right] = 0 \tag{7.93f}
$$

Equations (7.93d, e) may be uncoupled to yield

$$
\frac{u_{22}}{a} = \frac{v_{22}}{a} = \frac{\pi}{16} \left(\frac{w_{11}}{a} \right)^2 \tag{7.94}
$$

Then, using equations (7.94, 7.93a-c), we can rewrite the last of equations (7.93) as

$$\left(\frac{w_{11}}{a}\right)\left[\frac{\pi^2}{4}\left(\frac{w_{11}}{a}\right)^2 - \bar{u} + \frac{\pi^2}{3(1-v^2)}\left(\frac{h}{a}\right)^2\right] = 0 \tag{7.95}$$

Equation (7.95) has two solutions, that is

$$\frac{w_{11}}{a} = 0, \quad \left(\frac{w_{11}}{a}\right)^2 = \frac{4}{\pi^2}\left[\bar{u} - \frac{\pi^2}{3(1-v^2)}\left(\frac{h}{a}\right)^2\right] \tag{7.96}$$

Note that the second of these equations implies that a real solution can exist only if $\bar{u} \geq \bar{u}_{cr}$, where

$$\bar{u}_{cr} = \frac{\pi^2}{3(1-v^2)}\left(\frac{h}{a}\right)^2 \tag{7.97}$$

Thus, for $\bar{u} \leq \bar{u}_{cr}$ we have $\left(\dfrac{w_{11}}{a}\right) = 0$; for $\bar{u} > \bar{u}_{cr}$ we will have

$$\left(\frac{w_{11}}{a}\right)^2 = \frac{4}{\pi^2}(\bar{u} - \bar{u}_{cr}) \tag{7.98}$$

The critical point, then, at which buckling occurs, as reflected by a non-trivial value of the transverse deflection, is $\bar{u} = \bar{u}_{cr}$.

How does this result relate to our previous results for the buckling of rectangular (or square) plates? To answer this question, we first consider the *prebuckling state*, i.e., $\bar{u} < \bar{u}_{cr}$. Since w_{11} is zero here, it follows that $\bar{v} = v\bar{u}$ and that $u_{20} = v_{02} = u_{22} = v_{22} = 0$. Thus the only plate deformation, other than the contraction due to $\bar{u}$, is the Poisson expansion. Further, it is easily shown that the prebuckling stress is given by

$$\sigma_x = -E\bar{u} \tag{7.99}$$

When the (critical) *buckling state* is reached, $\bar{u} = \bar{u}_{cr}$. The other deflection components remain trivial, except $\bar{v} = v\bar{u}_{cr}$. Then

$$(\sigma_x)_{cr} = \frac{(N_x)_{cr}}{h} = -E\bar{u}_{cr} = -\frac{E\pi^2}{3(1-v^2)}\left(\frac{h}{a}\right)^2 \tag{7.100a}$$

or

$$(\sigma_x)_{cr} = -\frac{E\pi^2}{12(1-v^2)}\left(\frac{h}{a}\right)^2 \quad (4) \tag{7.100b}$$

Comparison of equations (7.100) and (7.16) yields a familiar result, $k_{cr} = 4$. Thus our enforced displacement yields the same buckling stress.

In the *postbuckling region*, the stress σ_x can be shown to be

$$\sigma_x = \frac{E\bar{u}_{cr}}{2}\left\{-2\frac{\bar{u}}{\bar{u}_{cr}} + \left(\frac{\bar{u}}{\bar{u}_{cr}} - 1\right)\left(1 - \cos\frac{2\pi y}{a}\right)\right\} \tag{7.101}$$

Equation (7.101) is obtained by using equations (7.91, 7.93, 7.94, 7.98) in

$$N_x = h\sigma_x = \frac{Eh}{1-v^2}\left[\frac{\partial u}{\partial x} + v\frac{\partial v}{\partial y} + \frac{1}{2}\left(\frac{\partial w}{\partial x}\right)^2 + \frac{v}{2}\left(\frac{\partial w}{\partial y}\right)^2\right] \tag{7.102}$$

It is also useful to calculate (from equation (7.101)) the average stress,

$$(\sigma_x)_{\text{ave}} = \frac{1}{a}\int_0^a \sigma_x\,dy = -\frac{E}{2}(\bar{u} + \bar{u}_{cr}) \tag{7.103}$$

Now we will plot (figure 3) equations (7.99) and (7.103). We see here a graphic display of the postbuckling load carrying capacity of the plate. Thus the plate continues to carry increased loading, beyond buckling, albeit at one-half of the rate prior to buckling. Remember, too, that this analysis assumes linear elastic material behavior throughout, so that plastic failure is excluded from consideration.

In figure 4 we have plotted the distribution of the axial stress σ_x as a function of the plate coordinate y, as given by equation (7.101). We note that the center of the plate holds at the same value $(\sigma_x)_{cr}$ even as the plate continues to be compressed, i.e., as $\bar{u}/\bar{u}_{cr}$ is increased. The additional load is carried by those portions of the plate near the edges. This phenomenon has given rise to the concept of the *effective width*, defined as

$$\frac{a_{\text{effective}}}{a} = \frac{(\sigma_x)_{\text{ave}}}{E\bar{u}} = \frac{1}{2}\left(1 + \frac{\bar{u}_{cr}}{\bar{u}}\right) = \frac{a^*}{a} \tag{7.104}$$

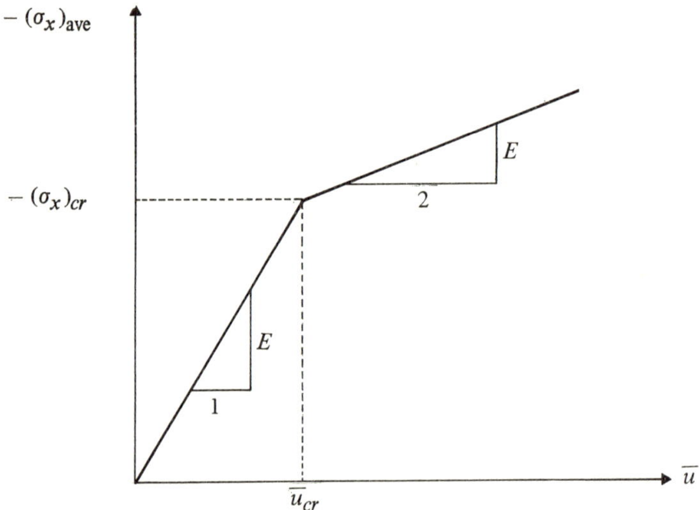

Fig. 3. Average stress and mean shortening for a buckled plate, after Mayers [8].

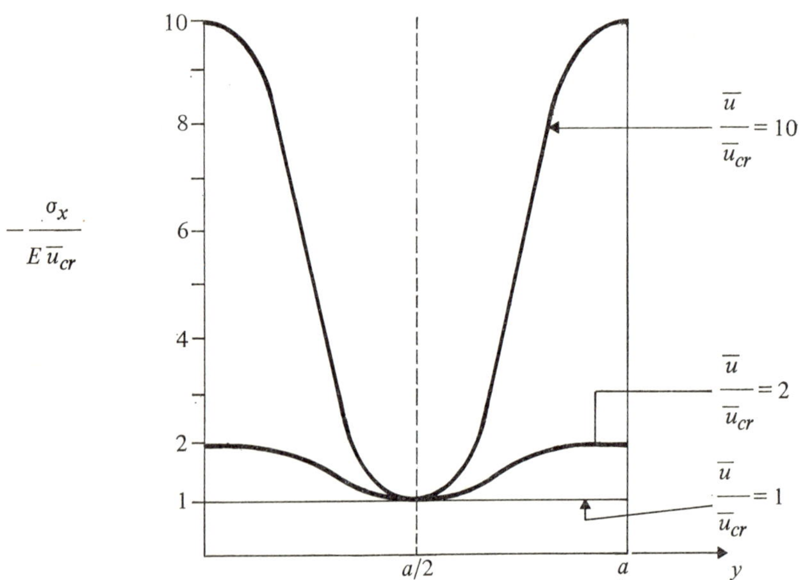

Fig. 4. Distribution of longitudinal stress of a buckled plate, after Mayers [8].

Physically this is the ratio of the average stress after buckling, for a given value of $\bar{u}$, to the stress that an unbuckled plate would carry at the same value of the shortening. The value of a^* thus denotes the fraction of the plate width that is actively carrying the applied load. At buckling, the load is uniformly distributed, and $a^* = a$. When $\bar{u}/\bar{u}_{cr} = 10$, little more than half the plate is effective (cf. figure 4), and $a^* = 0.55a$.

We also note that equation (7.98) can be written in the form

$$\frac{\bar{u}}{\bar{u}_{cr}} = 1 + \frac{3(1 - v^2)}{4}\left(\frac{w_{11}}{h}\right)^2 \tag{7.105}$$

which, by virtue of equations (7.100, 7.103), can be massaged into the form

$$\frac{(\sigma_x)_{\text{ave}}}{(\sigma_x)_{cr}} = 1 + \frac{3(1 - v^2)}{8}\left(\frac{w_{11}}{h}\right)^2 \tag{7.106}$$

Finally we will look at the problem of applying a given axial load, rather than prescribing the shortening of the plate. In this instance, with $\bar{N}_x$ positive in compression, we must add the following potential term to the strain energy (7.90),

$$V = \bar{N}_x \int_0^a \int_0^a \frac{\partial u}{\partial x}\,\mathrm{d}x\,\mathrm{d}y = -\bar{N}_x a^2 \bar{u} \tag{7.107}$$

Then, in addition to the differentiations carried out with respect to the other displacement coefficients, whose results do not change, we now must vary with respect to $\bar{u}$ as well. This is because $\bar{N}_x$ is prescribed now, not $\bar{u}$. Then minimizing the energy terms (7.92, 7.107) with respect to $\bar{u}$ yields the result

$$\bar{u} = \frac{\bar{N}_x}{Eh} + \frac{\pi^2}{8}\left(\frac{w_{11}}{a}\right)^2 \tag{7.108}$$

This result (remember that both $\bar{u}$, $\bar{N}_x$ are positive in compression) may be used to eliminate $\bar{u}$ from the second of equations (7.95). We then obtain

$$\frac{\pi^2}{8}\left(\frac{w_{11}}{a}\right)^2 = \left[\frac{\bar{N}_x}{Eh} - \frac{(\bar{N}_x)_{cr}}{Eh}\right] \tag{7.109}$$

where $(\overline{N}_x)_{cr}$ is the buckling load of the square plate given in equation (7.16). Then, again, transverse deflection is possible only if $\overline{N}_x$ exceeds the buckling load. Also, equation (7.109) can be rewritten as

$$\frac{\overline{N}_x}{(\overline{N}_x)_{cr}} = 1 + \frac{3(1 - v^2)}{8}\left(\frac{w_{11}}{h}\right)^2 \tag{7.110}$$

This, not surprisingly, indicates that in the previous problem we could identify $(N_x)_{ave}$ with the applied load $\overline{N}_x$ of this problem.

7.5   Summary

We have considered in this chapter the buckling and postbuckling behavior of rectangular and circular plates. The axisymmetric circular plate was used to demonstrate again the asymptotic analysis of early postbuckling behavior. The square plate analysis was conducted in an 'old fashioned' manner, and was used to introduce the concepts of effective width and of prescribed displacement buckling, as well as prescribed load.

We might point out that the Koiter theory, as we have developed it, in either formalism, could also have been applied to the rectangular plate. This is not especially surprising, although a very cursory analysis might have indicated that this was not the case. An implicit assumption of the Koiter theory is that there is a single buckling mode shape that corresponds to the buckling load. If there were more than one shape, the expansion would have to be in powers of ε_i, where i would range from unity to the number of mode shapes corresponding to the buckling load. Each shape thus requires its own expansion parameter. A casual interpretation of figure 1 would imply that such is the case for the rectangular plate under uni-axial compression. However, it is also clear from that figure that for a fixed geometry, i.e., a fixed plate aspect ratio a/b, there is but a single mode for which $k_{cr} = 4$.

References

[1] C. L. DYM and I. H. SHAMES: *Solid Mechanics: A Variational Approach.* McGraw-Hill Book Co., New York, 1973.

[2] F. B. HILDEBRAND: *Advanced Calculus for Applications.* Prentice-Hall, New Jersey, 1962.

[3] F. BOWMAN: *Introduction to Bessel Functions.* Dover Publications, New York, 1958.

[4] B. BUDIANSKY: Postbuckling Behavior of Cylinders in Torsion. In *Theory of Thin Shells*, edited by F. I. Niordson, Springer-Verlag, Berlin, 1969.

[5] J. M. T. THOMPSON and G. M. LEWIS: Continuum and Finite Element Branching Studies of the Circular Plate. *Computers and Structures*, Vol. 2, pp. 511-534, 1972.

[6] K. O. FRIEDRICHS and J. J. STOKER: Buckling of the Circular Plate Beyond the Critical Thrust. *Journal of Applied Mechanics*, Vol. 9, p. 7, 1942.

[7] J. MAYERS and B. BUDIANSKY: *Analysis of Behavior of Simply Supported Flat Plates Compressed Beyond the Buckling Load into the Plastic Range.* U.S. National Advisory Committee for Aeronautics Technical Note 3368, Langley Field, February 1955.

[8] J. MAYERS: *Aircraft and Missile Structures.* Lecture Notes, Stanford University, Stanford, California, 1964.

[9] S. P. TIMOSHENKO and J. M. GERE: *Theory of Elastic Stability.* McGraw-Hill Book Co., New York, 1961.

[10] R. M. RIVELLO: *Theory and Analysis of Flight Structures.* McGraw-Hill Book Co., New York, 1969.

8

Buckling and postbuckling of circular arches

8.1   The shallow arch—A closed-form solution

In this section we will present the highlights of a closed-form solution for the buckling of clamped, shallow arches, under 'dead' pressure, i.e., constant directional pressure. The solution was developed by Schreyer and Masur [1], although our summary presentation will differ in some details.

We begin with the nonlinear equations of a shallow arch in dimensionless form. Thus we write (see equations (5.57)) the equilibrium equations as

$$N' = 0, \quad -N - (N\chi)' - M'' = p \tag{8.1}$$

the constitutive relations as

$$N = \frac{1}{H} S = \frac{1}{H}(e + \tfrac{1}{2}\chi^2), \quad M = K \tag{8.2}$$

and the kinematic relations as

$$e = v' - w, \quad \chi = w', \quad K = -w'' \tag{8.3}$$

Here (and generally in this chapter) the prime is used to denote differentiation with respect to the dimensionless coordinate $\phi = y/R$, where R is the arch radius. The range of ϕ is such that $-\alpha \le \phi \le \alpha$. The tangential displacement, v, and the radial displacement, w, have also been rendered dimensionless with respect to the arch radius. The stress and moment resultants, respectively N and M, have been rendered dimensionless by dividing their

144

physical counterparts by (EI/R^2) and (EI/R), respectively. A thickness ratio, sometimes called a compressibility parameter, H, has also been introduced, i.e.,

$$H = \frac{1}{12}\left(\frac{h}{R}\right)^2 = \frac{I}{AR^2} \tag{8.4}$$

where h is the arch thickness, A is the cross-sectional area, and I the corresponding moment of inertia. Finally a dimensionless uniform pressure, p, has been defined as

$$p = \frac{qR^3}{EI} \tag{8.5}$$

Before proceeding with the analysis, we shall briefly indicate the nature of the results we seek. In figure 1 we have shown a schematic load-deflection curve, as a diagram of the pressure, p, vs. twice the average arch deflection, $\overline{w}$, which is

$$\overline{w} = \frac{1}{\alpha}\int_{-\alpha}^{\alpha} w \, d\phi \tag{8.6}$$

A variety of paths are possible, representing a variety of types of behavior. The chief variables are the rise of the arch, which is the height of the crown

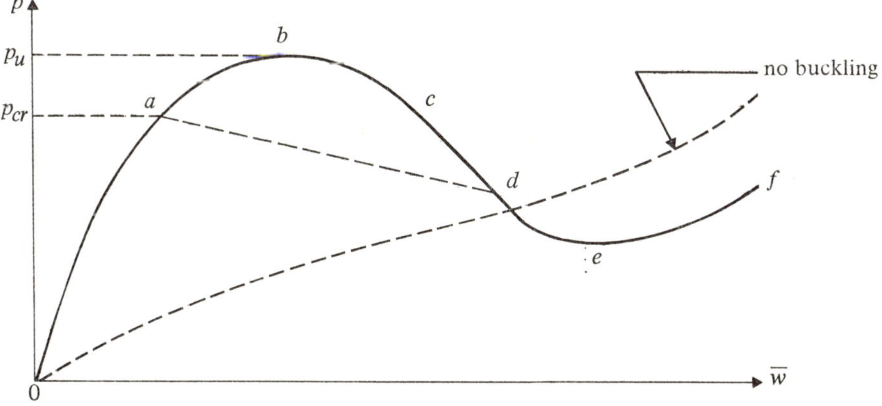

Fig. 1. Schematic load-deflection plot for an arch, after Schreyer and Masur [1].

145

($\phi = 0$) above the supports ($\phi = \pm \alpha$), and the pressure. If the arch rise is z, then

$$\frac{z}{R} = 1 - \cos \alpha \tag{8.7}$$

For small angles, which correspond to low arches, $\cos \alpha \simeq 1 - \alpha^2/2$, which allows the introduction of a dimensionless rise parameter, λ, defined as

$$\lambda = \frac{2z}{h} = \frac{\alpha^2 R}{h} \tag{8.8}$$

Then as λ increases, we can point to the following sequence of arch equilibrium paths:

1)   For very shallow arches (here, $\lambda < 2.85$) no buckling occurs, and we simply have a nonlinear spring, as indicated by the dotted curve in figure 1.
2)   For the range $2.85 \leq \lambda \leq 5.02$, only *symmetric, limit point* buckling occurs. This buckling is often called *snap buckling*, and it corresponds to the point '*b*' on the load-deflection curve. A condition for finding this limit point load, p_u, is $dp/d\bar{w} = 0$. Because $w(-\phi) = w(\phi)$, it is a symmetric buckling problem.
3)   For $5.02 \leq \lambda \leq 5.74$, *asymmetric bifurcations* occur, in the neighborhood of point '*c*'. These branch points thus occur after limit point buckling, that is, after the arch has followed the equilibrium path '*a-b-c*'. Thus the critical behavior of the arch is still represented by the limit point, p_u.
4)   For the region $\lambda \geq 5.74$, the controlling behavior is *asymmetric bifurcation* at point '*a*'. Note that the bifurcation occurs *before* the limit point, and that the arch will then deform so as to be on the branch '*a-d*' of the equilibrium diagram.

The above summary indicates the particular behavior that is of importance to us. The work of Schreyer and Masur is complete, however, and completes the curve through the point '*f*'. We are primarily interested in indicating the behavior summarized above, partly for its own sake and partly to provide a basis for comparison with the asymptotic analyses to be presented later.

We will first find the equations governing the symmetric equilibrium path, i.e., we will obtain the symmetric solution of equations (8.1). Let

$N = - N_0$ be the constant membrane force. Then from equations (8.1-8.3) we can find the governing equation as

$$w^{IV}(\eta) + w''(\eta) = \frac{\alpha^2}{\mu^2} q \tag{8.9}$$

where we have introduced

$$\eta = \phi\sqrt{N_0}, \quad \mu = \alpha\sqrt{N_0} \tag{8.10}$$

and

$$q = \frac{\alpha^2}{\mu^2} p - 1 \quad \text{or} \quad p = \frac{\mu^2}{\alpha^2}(q + 1) \tag{8.11}$$

Hence the parameter μ is a measure of the membrane force in the arch. It will also turn out to be a very useful parameter for systematically following the arch behavior.

A symmetric solution to equation (8.9) that satisfies the boundary conditions $w(\mu) = w'(\mu) = 0$ is

$$w(\eta) = \alpha^2 q\left[\frac{1}{2}\left(\frac{\eta^2}{\mu^2} - 1\right) + \frac{\cos\eta - \cos\mu}{\mu\sin\mu}\right] \tag{8.12}$$

The equilibrium path corresponding to this solution can be found from the relation

$$\int_{-\alpha}^{\alpha} S\,d\phi = - 2\alpha H N_0 \tag{8.13}$$

Noting equations (8.2, 8.3), and that $v(\pm\mu) = 0$, it follows that equation (8.13) can be written as

$$\int_{-\mu}^{\mu}\left[\frac{\mu^2}{2\alpha^2}(w'(\eta))^2 - w(\eta)\right]d\eta = - \frac{2\mu^3 H}{\alpha^2} \tag{8.14}$$

Then if we substitute our equilibrium solution (8.12) into the above integral and perform some algebra, we find the symmetric equilibrium path defined by the quadratic equation

$$Aq^2 + 4Bq - \frac{1}{3}\frac{\mu^4}{\lambda^2} = 0 \tag{8.15}$$

where

$$A = 4 - \tfrac{5}{3}\mu^2 - 3\mu \cot \mu - \mu^2 \cot^2 \mu$$

$$B = 1 - \tfrac{1}{3}\mu^2 - \mu \cot \mu \tag{8.16}$$

Note that we can also obtain here the average deflection, $\bar{w}$, and the deflection at the crown, w_0, i.e.,

$$\bar{w} = 2B\left(\frac{\alpha^2}{\mu^2}\right)q$$

$$w_0 = w(\eta = 0) = \alpha^2 q\left[\frac{1}{\mu}\tan\frac{\mu}{2} - \frac{1}{2}\right] \tag{8.17}$$

To determine asymmetric bifurcation points, or branching points, that allow divergence from the symmetric equilibrium path defined by equations (8.15, 8.16), we shall investigate the energy change from the symmetric path to an asymmetric buckled state. The symmetric total potential energy is (see equation (5.52))

$$\pi = \frac{1}{H}\int_{-\alpha}^{\alpha} S^2 \, d\phi + \int_{-\alpha}^{\alpha} K^2 \, d\phi - 2p\int_{-\alpha}^{\alpha} w \, d\phi \tag{8.18}$$

If ($^\wedge$) denotes an asymmetric buckling quantity, and if, for example,

$$w = w + \hat{w}$$

$$S = S + \hat{S} + \frac{1}{2}\hat{\chi}^2$$

$$\hat{S} = \hat{e} + \chi\hat{\chi} \tag{8.19}$$

then the change in the energy between the (symmetric) unbuckled path and the (asymmetric) buckled path is

$$\Delta\pi = \frac{1}{H}\int_{-\alpha}^{\alpha} \hat{S}^2 \, d\phi + \int_{-\alpha}^{\alpha} \hat{K}^2 \, d\phi + \frac{S}{H}\int_{-\alpha}^{\alpha} \hat{\chi}^2 \, d\phi \tag{8.20}$$

In writing down the above result we have retained only the quadratic terms in the potential energy, as we are considering only the (asymmetric) buckling state. We have also noted the consequences of that asymmetry by, for example, using the result

$$\int_{-\alpha}^{\alpha} K \hat{K} \, d\phi = 0$$

The equations governing the buckling state are found by setting $\delta^{(1)}[\Delta\pi] = 0$. We find

$$\hat{S} = -H\hat{N}_0 = \text{const.}$$

$$\hat{w}^{IV}(\phi) + N_0\hat{w}''(\phi) = -\hat{N}_0(1 + w''(\phi)) \tag{8.21}$$

For asymmetric behavior, $\hat{N}_0 = 0$, so that

$$\hat{w}^{IV}(\eta) + \hat{w}''(\eta) = 0 \tag{8.22}$$

The asymmetric solution to this equation that satisfies the boundary conditions requires that

$$\mu^c \cot \mu^c = 1, \quad \mu^c = 1.43\pi \tag{8.23}$$

from which it follows

$$q^c = -\frac{2}{5}[1 \mp \Gamma_c], \quad p^c = \frac{1}{5}\left(\frac{\mu^c}{\alpha}\right)^2 [3 \pm 2\Gamma_c] \tag{8.24}$$

where

$$\Gamma_c = \left[1 - \frac{5}{4}\left(\frac{\mu^c}{\lambda}\right)^2\right]^{1/2} \simeq 1 - \frac{5}{8}\left(\frac{\mu^c}{\lambda}\right)^2 \tag{8.25}$$

We see from the first of equations (8.25) that a real solution requires that

$$\lambda > (\tfrac{5}{4})^{1/2}\mu^c = 5.02 \tag{8.26}$$

Thus, equation (8.26) gives the minimum requirement on the rise parameter λ for asymmetric bifurcation to take place.

The (symmetric) limit point buckling pressures are determined by the condition

$$\frac{dp}{d\mu} = 0 \tag{8.27}$$

which is applied in conjunction with equations (8.11, 8.15, 8.16). The result is again a quadratic equation for the load parameter, q,

$$Cq^2 + 2Aq + 2B + \frac{\mu^4}{6\lambda^2} = 0 \tag{8.28}$$

where A, B are as given in equations (8.16), and

$$C = 6 - \tfrac{29}{12}\mu^2 - \tfrac{15}{4}\mu\cot\mu - \tfrac{7}{4}\mu^2\cot^2\mu - \tfrac{1}{2}\mu^3\cot\mu - \tfrac{1}{2}\mu^3\cot^3\mu \tag{8.29}$$

Equations (8.28) and (8.15) are to be solved simultaneously to find those values of p, μ for which the condition (8.27) is satisfied. This can be done graphically, by plotting p vs. μ for different values of λ. Actually, at this point, it is not very important that we complete this particular calculation, as will be evident below.

We can now piece together the principal components of the arch behavior by following the parameter, μ, which you will recall is also proportional to the membrane force N_0. First of all, since for small values of μ [2]

$$\cot\mu \cong \frac{1}{\mu} - \frac{\mu}{3} - \frac{\mu^3}{45} - \frac{2\mu^5}{945} \tag{8.30}$$

we can easily find that

$$A \cong -\frac{4\mu^6}{945}, \quad B \cong \frac{\mu^4}{45} \tag{8.31}$$

Then the equilibrium paths will behave as (equation (8.15))

$$\mu^2 q \cong -\frac{21}{2}\left[-1 \pm \left(1 - \frac{5}{7}\frac{\mu^2}{\lambda^2}\right)\right] \tag{8.32}$$

150

so that by equation (8.11)

$$\lim_{\mu \to 0} p = 0, \quad \frac{21}{\alpha^2} \tag{8.33}$$

Thus the vanishing of the parameter μ yields two points on the schematic of figure 1, the origin and the point 'f'.

As we increase μ from zero, it appears that there will be singular behavior as μ approaches π. In fact, for μ very close to π,

$$A \cong \frac{-1}{\pi^2(\pi - \mu)^2}, \quad B \cong \frac{1}{\pi(\pi - \mu)} \tag{8.34}$$

We then obtain that, for μ in the neighborhood of π,

$$q \cong 2\pi(\pi - \mu)\left[1 \mp \sqrt{1 - \frac{\pi^4}{12\lambda^2}}\right] \tag{8.35}$$

In order for the radical to yield a real value, it must be true that

$$\lambda > \pi^2/\sqrt{12} = 2.85 \tag{8.36}$$

Equation (8.36) is a restriction on values of the rise parameter in order that the value of q (and consequently p, w_0, $\bar{w}$) be real when $\mu = \pi$. If this is not the case, i.e., if $\lambda < 2.85$, then μ never gets up to the value $\mu = \pi$. This is clear because equations (8.31) and (8.34) indicate that $-A$ and B are increasing from $\mu = 0$ to $\mu = \pi$. And it is important that μ reach, and exceed, the value π because this is the point beyond which multiple values of μ are possible (see equations (8.35), (8.11)) for a given value of p. In fact, for bifurcation, $\mu = \mu^c = 1.43\pi$.

Thus we can say that for $\lambda < 2.85$, two discrete (and different) values of p are obtained for each μ, and $\mu < \pi$ always. And it is also true, in this range, that each value of p occurs only once, that is, for a given value of λ there will be only one value of μ which can yield a particular value of p. This is exemplified by figure 2, which is figure 4 of Schreyer and Masur [1]. The graphs are simply plots of the roots of equation (8.15) for different values of the rise parameter.

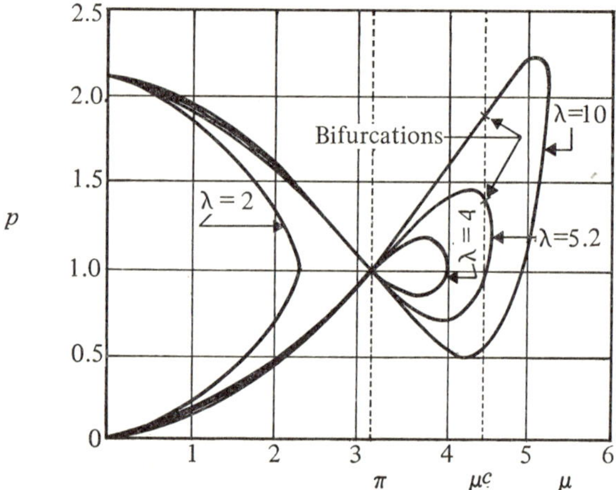

Fig. 2. Loading parameter plots for nonlinear arch behavior, after Schreyer and Masur [1].

We have also seen that asymmetric bifurcation requires that $\mu = \mu^c = 1.43\pi$, and $\lambda > 5.02$. And, for $\lambda < 2.85$ (and $\mu < \pi$), no buckling occurs. For $2.85 \leq \lambda \leq 5.02$, or $\pi \leq \mu \leq 1.43\pi$, what happens?

If we look at the curve labeled $\lambda = 4$, in figure 2, and follow it from the origin ($\mu = p = 0$) with increasing p, μ, we see that we reach a point where $dp/d\mu = 0$ long before μ attains the value $\mu = \mu^c$. Thus we would expect limit point buckling. This is also true for $\lambda = 5.02$ which is a value for which we expect bifurcation by the criterion (8.26). Indeed, shortly after the limit point is reached, the path crosses the line $\mu = \mu^c$. Thus the bifurcation occurs *after* the limit point, for $\lambda = 5.02$. For $\lambda = 10$, the bifurcation occurs well before the limit point is reached.

Thus, while a bifurcation point occurs whenever $\lambda > 5.02$, it is not clear whether this bifurcation will occur before or after the limit point. By solving equation (8.15) for the equilibrium path simultaneously with equation (8.28) for the limit point, with $\mu = \mu^c = 1.43\pi$, we are led to roots of $q = -0.21$ and -0.86, which in turn yield two values of λ. The first is $\lambda = 5.74$, and the second is a value less than $\lambda = 5.02$, for which bifurcation simply cannot occur. Thus at $\lambda = 5.74$, the limit point occurs simultaneously with $\mu = \mu^c$, the bifurcation point. By example, then, (figure 2) we have seen that for $\lambda > 5.74$ bifurcation precedes the limit point, and for $5.02 < \lambda < 5.74$ the bifurcation occurs after the limit point. Thus, in the latter case, limit point buckling should be our criterion.

Finally, one can verify by (numerically) solving equation (8.28), that for $2.85 < \lambda < 5.74$ a limit point can always be found, with $\mu \leq \mu^c$.

Thus, as we had planned, we have observed the following behavior, i.e.,

$0 < \lambda < 2.85$: no buckling

$2.85 < \lambda < 5.02$: limit point — no bifurcation

$5.02 < \lambda < 5.74$: bifurcation following limit point

$5.74 < \lambda$: bifurcation before limit point

The bifurcation pressure (8.24) can be approximated as, for large λ^2,

$$p^c \cong \left(\frac{\mu^c}{\alpha} \right)^2 \left[1 - \frac{(\mu^c)^2}{4\lambda^2} \right] \tag{8.37}$$

Although Schreyer and Masur [1] did not deal with the *pinned* arch under dead pressure, it is not difficult to do that analysis to find

$0 < \lambda < 1.11$: no buckling

$1.11 < \lambda < 2.24$: limit point — no bifurcation

$2.24 < \lambda < 2.65$: bifurcation following limit point

$2.65 < \lambda$: bifurcation before limit point

The bifurcation pressure is such that

$$p^p \cong \left(\frac{\pi}{\alpha} \right)^2 \left[1 - \frac{1.89}{\lambda^2} \right] \tag{8.38}$$

We might note that Dickie and Broughton [3] have given extensive numerical results for the transition points and buckling pressures for a variety of loads and boundary conditions. Their analysis was done in a rather different manner, although for the same cases they achieved the same results.

Before closing this section it is worth noting two points. The first relates to the path followed by the arch, on the schematic diagram in figure 1, after a bifurcation at point 'a'. It is quite possible, depending on the loading device, for the load to fall off after bifurcation. Then the arch would follow the path '*a-d*'. This would be true for a loading device that really controlled the displacement, as, say, a typical large scale testing machine. For true dead loading, e.g., weights, the load would remain constant after bifurcation, and the arch would *snap* along a horizontal to a point on the curve '*e-f*'. This is basically a kinetic process, where in practice a loud snap is heard due to the release of kinetic energy.

The remaining item is simply to point out some orders of magnitude for the rise parameter λ. We note from equations (8.7) and (8.8) that while the arch semi-vertex angle, α, in radians, is small for a shallow arch, the rise parameter is proportional to the ratio of the radius of curvature of the arch to its thickness. One definition of a shallow arch is to restrict its rise to be less than one eighth of its subtended chord [4]. This restriction, together with the small angle assumption, implies that $\alpha < 1/2$ (or $\alpha° < 30°$). For a comparatively thick arch, say $R/h = 20$, $\lambda = 5$; while for a thin arch of $R/h = 100$, say, $\lambda = 25$. The point of this simple speculation is that for thin arches, the rise would have to be very small for bifurcation not to be the governing buckling criterion.

8.2 Asymptotic analysis for arches

We will derive in this section the general formulae for the postbuckling analysis, according to the Budiansky-Hutchinson formalism, of arches both shallow and steep. We will follow the analysis of Dym [5].

Before proceeding with the postbuckling analysis, we observe that if we use the notation of equations (8.2), the formalism of the postbuckling analysis can proceed for both arch theories developed in Section 5.3. Thus in addition to equations (8.1) and (83.) for shallow arches, we need only identify corresponding results for the steep arch theory. From equations (5.54) we can find the dimensionless equilibrium equations

$$N' - N\chi - M' = 0, \quad N + (N\chi)' + M'' + p = 0 \tag{8.39}$$

while from equation (5.49) we have the kinematic relations

$$e = v' - w, \quad \chi = w' + v, \quad K = -(w'' + v') \tag{8.40}$$

Then, for either set of arch equations, we can easily see that a common virtual work statement for the uniform pressure loading of an arch is

$$\int_{-\alpha}^{\alpha} [N\delta e + N\chi\delta\chi + M\delta K - p\delta w]\,d\phi = 0 \tag{8.41}$$

We now introduce the following expansions for the non-varied quantities above,

$$v = pv_0 + \varepsilon v_1 + e^2 v_2 + \ldots$$

$$w = pw_0 + \varepsilon w_1 + \varepsilon^2 w_2 + \ldots$$

$$e = pe_0 + \varepsilon e_1 + \varepsilon^2 \varepsilon_2 + \ldots$$

$$\chi = \varepsilon\chi_1 + \varepsilon^2\chi_2 + \ldots$$

$$K = pK_0 + \varepsilon K_1 + \varepsilon^2 K_2 + \ldots$$

$$N = -N_0 + \varepsilon N_1 + \varepsilon^2 N_2 + \ldots$$

$$M = M_0 + \varepsilon M_1 + \varepsilon^2 M_2 + \ldots \tag{8.42}$$

As usual ε is a small parameter such that $p \to p_{cr}$ when $\varepsilon \to 0$. The quantity χ_0 was deleted from the expansion for the rotation, χ, so that the prebuckling state would be a linear state. If we substitute the expansions (8.42) into the variational statement (8.41) we find

$$\int_{-\alpha}^{\alpha} [-N_0\delta e + M_0\delta K - p\delta w]\,d\phi$$

$$+ \varepsilon \int_{-\alpha}^{\alpha} [N_1\,\delta e + M_1\,\delta K - N_0\chi_1\,\delta\chi]\,d\phi$$

$$+ \varepsilon^2 \int_{-\alpha}^{\alpha} [N_2\,\delta e + M_2\,\delta K + (N_1\chi_1 + N_0\chi_2)\delta\chi]\,d\phi$$

$$+ \varepsilon^3 \int_{-\alpha}^{\alpha} [N_3\,\delta e + M_3\,\delta K + (N_1\chi_2 + N_2\chi_1 - N_0\chi_3)\delta\chi]\,d\phi$$

$$+ \ldots = 0 \tag{8.43}$$

The vanishing of the first term in the expanded variational statement (8.43) defines the prebuckling problem, i.e.,

8 *Buckling and postbuckling of circular arches*

$$\int_{-\alpha}^{\alpha} [-N_0 \, \delta e + M_0 \, \delta K - p \, \delta w] \mathrm{d}\phi = 0 \tag{8.44}$$

Then an ε can be divided through in equation (8.43), so that when we pass to the limit $\varepsilon = 0$ we define the buckling problem as

$$\int_{-\alpha}^{\alpha} [N_1 \, \delta e + M_1 \, \delta K - \psi p_{cr}\chi_1 \, \delta\chi] \mathrm{d}\phi = 0 \tag{8.45}$$

where ψ is a parameter that will be used later in the prebuckling problem, i.e., we shall set $N_0 = p\psi$.

Since the displacement and strain variations in the above variational statements need only be kinematically consistent, we can certainly replace them with their buckling counterparts. From equation (8.45) we would thus obtain the Rayleigh quotient for the buckling pressure

$$\psi p_{cr} = \frac{\displaystyle\int_{-\alpha}^{\alpha} [N_1 e_1 + M_1 K_1] \mathrm{d}\phi}{\displaystyle\int_{-\alpha}^{\alpha} \chi_1^2 \, \mathrm{d}\phi} \tag{8.46}$$

If we make the same substitution in equation (8.43), while noting the above result, we obtain

$$(p_{cr} - p)\psi \int_{-\alpha}^{\alpha} \chi_1^2 \, \mathrm{d}\phi$$

$$+ \, \varepsilon \int_{-\alpha}^{\alpha} [N_2 e_1 + M_2 K_1 + N_1\chi_1^2 - N_0\chi_1\chi_2] \mathrm{d}\phi$$

$$+ \, \varepsilon^2 \int_{-\alpha}^{\alpha} [N_3 e_1 + M_3 K_1 + N_1\chi_1\chi_2 + N_2\chi_1^2 - N_0\chi_1\chi_3] \mathrm{d}\phi$$

$$+ \ldots = 0 \tag{8.47}$$

As it would appear from equation (8.47) that we would need to solve the third order problem to obtain the coefficient of ε^2 in the load expansion, we could conveniently introduce here orthogonality conditions on the higher order terms to circumvent that problem. We choose to require that

156

$$\int_{-\alpha}^{\alpha} [N_1 e_m + M_1 K_m] \mathrm{d}\phi = 0, \quad m = 2, 3, 4 \ldots \tag{8.48}$$

It follows from equations (8.48) and (8.45), since the mth order mode shapes must also be kinematically admissible, that

$$\int_{-\alpha}^{\alpha} \chi_1 \chi_m \, \mathrm{d}\phi = 0, \quad m = 2, 3, 4 \ldots \tag{8.49}$$

In order to apply the orthogonality condition (8.48) and its consequence (8.49) successfully, we note the following consequences of constitutive linearity and geometric nonlinearity, i.e.,

$$N_2 e_1 = N_2 S_1 = N_1 S_2 = N_1(e_2 + \tfrac{1}{2}\chi_1^2)$$

$$N_3 e_1 = N_3 S_1 = N_1 S_3 = N_1(e_3 + \chi_1 \chi_2)$$

$$M_2 K_1 = M_1 K_2; \quad M_3 K_1 = M_1 K_3 \tag{8.50}$$

In equations (8.50) we have used the notation that S_i is the total nonlinear strain of order ε^i, which is made up in part of products of lower order terms due to the geometric nonlinearity. Then with the aid of equations (8.50), (8.48) and (8.49), we can obtain finally the desired expansion for the post-buckling pressure. Thus we have

$$\frac{p}{p_{cr}} = 1 + a\varepsilon + b\varepsilon^2 + \ldots \tag{8.51}$$

where

$$a = \frac{3}{2\psi p_{cr}} \frac{\displaystyle\int_{-\alpha}^{\alpha} [N_1 \chi_1^2] \mathrm{d}\phi}{\displaystyle\int_{-\alpha}^{\alpha} \chi_1^2 \, \mathrm{d}\phi} \tag{8.52a}$$

$$b = \frac{1}{\psi p_{cr}} \frac{\displaystyle\int_{-\alpha}^{\alpha} [2 N_1 \chi_1 \chi_2 + N_2 \chi_1^2] \mathrm{d}\phi}{\displaystyle\int_{-\alpha}^{\alpha} \chi_1^2 \mathrm{d}\phi} \tag{8.52b}$$

157

Recall that according to this theory, if $a \neq 0$, then the structure can be in an unstable state in the postbuckling range, and it will also be imperfection sensitive. If $a = 0, b \neq 0$, then the postbuckling stability and the imperfection sensitivity vary as the arithmetic sign of b.

In the present result it appears at first glance that the existence of a stable or unstable equilibrium state in the postbuckling range depends on whether or not inextensibility is assumed in the buckling state. For, from equation (8.52), $a \neq 0$ if $N_1(\phi) = H^{-1} e_1(\phi)$ is not zero. However, it seems rather clear that for asymmetric buckling, the membrane stress $N_1(\phi)$ must itself be asymmetric. (Indeed it may even be zero; remember that we found $\hat{N}_0 = 0$ in the previous section.) It may also be argued that the pressure-loaded arch is a symmetrical structure which bifurcates asymmetrically, so that $a = 0$.

Before proceeding to the evaluation of the coefficient b, in the next section, we note that the approach of Koiter applied to this problem would yield, for the potential energy at buckling and in the postbuckling range, the following [6]:

$$P^p = \left\{ \psi p_{cr} \int_{-\alpha}^{\alpha} \chi_1^2 \, \mathrm{d}\phi \right\} \left[-\left(\frac{p}{p_{cr}} - 1 \right) \varepsilon^2 + \tfrac{2}{3} a \varepsilon^3 + \tfrac{1}{2} b \varepsilon^4 \right] \quad (8.53)$$

The load-deflection relation (8.51) can be obtained from equation (8.53) by application of the condition

$$\frac{\mathrm{d}P^p}{\mathrm{d}\varepsilon} = 0 \quad\quad (8.54)$$

Further, the energy at the critical load is

$$P^p \big|_{p = p_{cr}} = \left\{ \psi p_{cr} \int_{-\alpha}^{\alpha} \chi_1^2 \, \mathrm{d}\phi \right\} [\tfrac{2}{3} a \varepsilon^3 + \tfrac{1}{2} b \varepsilon^4] \quad\quad (8.55)$$

Then the application of the energy criteria to this problem is clear cut. The quadratic functional corresponding to the buckling energy is the term of order ε^2 in equation (8.53). It (or its second variation) ceases to be positive definite when $p = p_{cr}$. For stability at and immediately beyond the bifurcation point, equation (8.55) implies that the sufficient condition is $a = 0$, $b > 0$, which is, of course, a familiar result.

8.3   Shallow arch results

In this section we shall outline briefly the results of the asymptotic analysis for shallow arches, both clamped and pinned. For the clamped arches we will give some of the details; for the latter we will just outline the results.

The equations that govern prebuckling, buckling, and initial post-buckling states can be found by perturbing the differential equations (8.1), or by the vanishing of the variational statement (8.43) with $p \rightarrow p_{cr}$ as $\varepsilon \rightarrow 0$. In either case we have

$$0(1) \qquad N_0' = 0, \qquad w_0^{IV} = 1 - \psi \tag{8.56}$$

$$0(\varepsilon) \qquad N_1' = 0, \qquad w_1^{IV} + \psi p_{cr} w_1'' = N_1 \tag{8.57}$$

$$0(\varepsilon^2) \quad N_2' = 0, \qquad w_2^{IV} + \psi p_{cr} w_2'' = N_2 + N_1 w_1'' \tag{8.58}$$

where ψ is such that $N_0 = p\psi$. Remember that the prebuckling state will be symmetric, the buckling state asymmetric, while the second order state is even. We also note that the constant membrane forces N_0, N_1, N_2, can be found by perturbing the condition

$$\int_{-\alpha}^{\alpha} v' \, d\phi = 0 = 2NH\alpha + \int_{-\alpha}^{\alpha} [w - \tfrac{1}{2}\chi^2] \, d\phi \tag{8.59}$$

For the clamped arch, the solution to equation (8.56) is easily found to be

$$w_0 = \frac{1}{24} (1 - \psi) (\alpha^2 - \phi^2)^2 \tag{8.60}$$

Then, from the zeroth order version of equation (8.59) we can find that

$$\psi_c = \left(1 + \frac{45H}{\alpha^4}\right)^{-1} = \left(1 + \frac{15}{4\lambda^2}\right)^{-1} \tag{8.61}$$

For asymmetric buckling, $w_1(-\phi) = -w_1(\phi)$, and thus from the first order version of equation (8.59), it follows that $N_1 = 0$. Thus equation (8.57) is now reduced (appropriately) to an eigenvalue problem, for which

$$w_1(\phi) \sim \left[\frac{\phi}{\alpha} - \frac{\sin \phi \sqrt{\psi p_{cr}}}{\sin \alpha \sqrt{\psi p_{cr}}}\right] \tag{8.62}$$

where the eigenvalues are determined as the roots of

$$\alpha\sqrt{p_{cr}}\psi = \tan(\alpha\sqrt{p_{cr}}\psi)$$

which are already known to us, i.e.,

$$(p_{cr})_c = \left(\frac{1.43\pi}{\alpha}\right)^2\left(1 + \frac{15}{4\lambda^2}\right) \tag{8.63}$$

We will assume also that $w_1(\phi)$ is properly normalized so that its maximum deflection is equal to the arch thickness.

Before proceeding further, we note that with $N_1 = 0$, and $N_2' = 0$, the coefficient we wish to calculate is simply

$$b = \frac{N_2}{\psi p_{cr}} \tag{8.64}$$

The stress resultant N_2 is obtained by simultaneous solution of equation (8.58) and (from equation (8.59))

$$2\alpha H N_2 + \int_{-\alpha}^{\alpha}\left[w_2 - \tfrac{1}{2}\chi_1^2\right]\mathrm{d}\phi = 0 \tag{8.65}$$

The result of this calculation is such that

$$b_c = \frac{1.61}{1 - \dfrac{\lambda^2}{5.02}} \tag{8.66}$$

Thus we see that whenever $\lambda^2 > 5.02$, the clamped arch is unstable. This is clearly in consonance with our exact results, given in the previous section. We also note that if we had normalized the buckling deflection with respect to the radius, rather than the thickness, we could write equation (8.66) as

$$b_c = -\frac{(1.61)\,(5.02)}{\alpha^4\left[1 - \dfrac{5.02}{\lambda^2}\right]} \tag{8.67}$$

For the pinned arch we can find that

$$(p_{cr})_p = \left(\frac{\pi}{\alpha}\right)^2 \left(1 + \frac{5}{8\lambda^2}\right) \tag{8.68}$$

and

$$b_p = -\frac{3(1.90)}{\alpha^4 \left[1 - \dfrac{1.90}{\lambda^2}\right]} \tag{8.69}$$

Thus the pinned arch analysis yields a bifurcation pressure reasonably close to the exact result, as did the clamped arch analysis. Further, the pinned arch is predicted to be unstable whenever $\lambda^2 > 1.90$, which is in reasonable agreement with our exact results.

8.4   Steep arch results

For steeper arches we will present the results of Dym [5], without all the details. The principles are identical, but the algebra is considerably more complex. The governing differential equations are

$$0(1) \qquad N_0 = p \tag{8.70}$$

$$0(\varepsilon) \qquad N_1' + p_{cr}\chi_1 - M_1' = 0$$
$$M_1'' + N_1 - p_{cr}\chi_1' = 0 \tag{8.71}$$

$$0(\varepsilon^2) \qquad N_2' + p_{cr}\chi_2 - M_2' = N_1\chi_1$$
$$M_2'' + N_2 - p_{cr}\chi_2' = -(N_1\chi_1)' \tag{8.72}$$

The asymmetric buckling solution, for example, appears as

$$w_1 = A_1 \sin \mu\phi + A_2 \sin\phi + A_3 \phi \cos\phi$$

$$v_1 = -\frac{1}{\mu} A_1 \cos\mu\phi + (QA_3 - A_2)\cos\phi + A_3 \phi \sin\phi \tag{8.73}$$

where

$$\mu^2 = p, \quad Q = \frac{1 + H(\mu^2 - 1)}{1 - H(\mu^2 - 1)} \tag{8.74}$$

The buckling eigenvalues are then determined as the roots of the following transcendental equations, i.e., for the pinned arch

$$\mu \tan \mu\alpha = \frac{(1 + Q) \sin^2 \alpha}{(1 + 2Q - \mu^2 Q) \sin \alpha \cos \alpha + \alpha(1 - \mu^2)} \tag{8.75}$$

while for the clamped arch

$$\mu \tan \mu\alpha = \frac{(1 + \mu^2 Q) \sin \alpha \cos \alpha - \alpha(1 - \mu^2)}{(1 + Q) \cos^2 \alpha} \tag{8.76}$$

The constants A_i above are determined from two of the homogeneous equations leading to the respective eigenvalue problems and by stipulating that the maximum (radial) buckling displacement is equal to the radius of curvature of the arch.

The second order solution would be of the form

$$w_2 = D_1 \cos \mu\phi + D_2 \cos \phi + D_3 \phi \sin \phi + C_0 + \sum_{i=1}^{4} C_i \cos q_i \phi$$

$$v_2 = \frac{1}{\mu} D_1 \sin \mu\phi + (D_2 + QD_3) \sin \phi$$

$$- D_3 \phi \cos \phi + \sum_{i=1}^{4} C_{i+4} \sin q_i\phi \tag{8.77}$$

where $q_1 = 2$, $q_2 = 2\mu$, $q_{3,4} = 1 \pm \mu$. The C_k are determined for the particular solution of equations (8.72), and the D_k by then satisfying the boundary conditions.

It is useful to point out the following order of magnitude relations which can be discerned from the analytical results:

$$e_1 = 0(H), \quad \chi_1 = 0(1), \quad \chi_2 = 0(1)$$

$$e_2 = 0(1), \quad e_2 + \tfrac{1}{2}\chi_1^2 = 0(H) \tag{8.78}$$

In the tables that follow, we present some numerical results for various values of the semi-vertex angle α and the compressibility H. We also present some comparisons of results obtained from both shallow and steep arch theories. In Ref. [5] comparisons are made with still a third theory, to investigate another technical detail.

Table I   Bifurcation pressures for steep arches

	pinned		p_{cr} *clamped*	
α	$h/R = 1/10$	$h/R = 1/1000$	$h/R = 1/10$	$h/R = 1/1000$
90°	3.26923	3.27124	9.00000	9.00000
80°	4.51287	4.51449	11.33095	11.33135
70°	6.21586	6.21707	14.61636	14.61805
60°	8.72629	8.72712	19.58254	19.58671
50°	12.78071	12.78124	27.73864	27.74726
40°	20.14096	20.14128	42.68493	42.70190

In Table I we see that the critical pressures are sensitive to the steepness of the arch, as reflected in changes of the angle α, and to the boundary conditions, with the clamped arch being stiffer. However, the pressures at bifurcation are almost completely insensitive to the thickness-to-radius ratio, and thus to the extensibility of the arch centerline. Also, we note that as $\alpha \to \pi/2$, we obtain the critical pressure of a complete ring under dead pressure, from the pinned arch results.

Table II   Postbuckling coefficients for steep arches

	pinned		b *clamped*	
α	$h/R = 1/10$	$h/R = 1/1000$	$h/R = 1/10$	$h/R = 1/1000$
90°	1.019302	1.011415	-0.575536	-0.571700
80°	0.386367	0.381016	-1.204257	-1.190227
70°	-0.484773	-0.484626	-2.446248	-2.396215
60°	-2.210659	-2.184584	-5.194701	-4.993075
50°	-6.556133	-6.370323	-12.307776	-11.294352
40°	-20.298221	-18.804925	-36.999547	-29.413727

From the results in Table II we deduce that the postbuckling coefficients are also relatively insensitive to the extensibility, but they are very sensitive to the value of the semi-vertex angle. For the clamped arch, the coefficient is always negative, indicating unstable postbuckling behavior and imperfection sensitivity for all values of the rise. For the pinned arch, as α increases to $\alpha = \pi/2$ we approach the stable postbuckling behavior of a complete ring! But for shallower arches, the bifurcation is always followed by unstable postbuckling behavior.

In the last two tables are indicated comparative figures obtained from the leading terms of the shallow arch analyses and those given by our steep arch analysis. Here as α increases, the discrepancies increase, which is not too surprising. We should note that the more refined theory presented earlier in this section assumed that the pre-buckling state was a uniform membrane

Table III   Comparison of critical loads

	pinned		p_{cr}	*clamped*	
α	*steep*	*shallow*		*steep*	*shallow*
90°	3.27	4.00		9.00	8.19
80°	4.51	5.06		11.33	10.35
70°	6.22	6.61		14.62	13.52
60°	8.73	9.00		19.59	18.40
50°	12.78	12.96		27.75	26.55
40°	20.14	20.25		42.70	41.45

Table IV   Comparison of postbuckling coefficients

	pinned		b	*clamped*	
α	*steep*	*shallow*		*steep*	*shallow*
90°	+1.011	−0.933		−0.572	−1.33
80°	+ .381	−1.50		−1.190	−2.14
70°	−0.485	−2.54		−2.396	−3.72
60°	−2.184	−4.71		−4.993	−6.71
50°	−6.370	−9.80		−11.294	−13.92
40°	−18.805	−24.0		−29.414	−34.2

state. If bending was allowed in the pre-buckling state (see Ref. [5]), then the agreement for angles in the range $40° < \alpha < 60°$ would be even better, as far as the prediction of the postbuckling coefficients is concerned.

8.5 Summary

We have presented in this chapter a discussion of the buckling and post-buckling behavior of arches under dead pressure. We have used a closed-form solution from shallow arch theory to indicate the ranges of possible behavior. Then we presented a number of asymptotic analyses of bifurcation and initial postbuckling behavior, and indicated ranges of agreement. We were also able to touch briefly on the behavior of complete rings.

References

[1] H. L. Schreyer and E. F. Masur: Buckling of Shallow Arches. *Journal of the Engineering Mechanics Division*, ASCE, Vol. 92, No. EM4, August 1966, p. 1.

[2] M. Abramowitz and I. A. Stegun: *Handbook of Mathematical Functions.* Government Printing Office, Washington, D.C., 1965.

[3] J. F. Dickie and P. Broughton: Stability Criteria for Shallow Arches. *Journal of the Engineering Mechanics Division*, ASCE, Vol. 97, No. EM3, June 1971, p. 951.

[4] C. L. Dym: On Extensibility and the Buckling of Shallow and Steep Arches. *Proceedings of the Sixth Southeastern Conference on Theoretical and Applied Mechanics*, Tampa, Florida, March 1972.

[5] C. L. Dym: Buckling and Postbuckling Behavior of Steep Compressible Arches *International Journal of Solids and Structures*, Vol. 9, No. 1, January 1973, p. 129.

[6] C. L. Dym: Bifurcation Analyses for Shallow Arches. *Journal of the Engineering Mechanics Division*, ASCE, Vol. 99, No. EM2, April 1973, p. 287.

9

Some applications of Lyapunov functionals

In Chapter 4 we indicated briefly the scope of the Lyapunov functional methodology for analyzing the stability of continuous systems. In this chapter we shall display some applications chosen from the current research literature. The available literature includes the pioneering work of Zubov and Movchan [1-3]; applications to existence and uniqueness for linear elasticity theory [4, 5]; the very complete exposition and collection of examples given by Plaut [6]; applications to problems of aeroelasticity [7-9], structural stability [10-12], chemical engineering [13] and hydrodynamics [14]; applications to bounding procedures [15-19]; and the development and application of a methodology for constructing Lyapunov functionals for nonconservative linear (and linearized) elastic systems [20-23].

The examples to be displayed here were chosen in part for their simplicity, so that we would not become bogged down in computational detail, as well as for their wide variety of application. And we shall be examining in many instances the stability of solutions of partial differential equations, in addition to our usual interest in determining various critical values of loading or other parameters.

We also recall here, from Chapter 4, that to define the stability of a solution in a continuous system, we determine stability with respect to a *metric*, i.e., a functional representation of distance from the equilibrium state. Further, for stability, we need to construct a *Lyapunov functional V* such that 1) V is positive definite with respect to the metric; 2) V can be made to admit an infinitely small upper bound with respect to the metric; and 3) V is either zero (for stability) or negative definite (for asymptotic stability).

9 *Some applications of Lyapunov functionals*

9.1 The wave equation

Consider first the stability of the solution $u(x, t) = 0$, i.e., the equilibrium position, of the one-dimensional wave equation [4]

$$\frac{\partial^2 u}{\partial x^2} = \frac{\partial^2 u}{\partial t^2} \tag{9.1}$$

subject to the boundary conditions, say,

$$u(0) = u(1) = 0 \tag{9.2}$$

For a metric we choose

$$\rho^2 = \int_0^1 \left[\left(\frac{\partial u}{\partial t} \right)^2 + u^2 + \left(\frac{\partial u}{\partial x} \right)^2 \right] dx \tag{9.3}$$

and as a Lyapunov functional we take twice the total energy, i.e.,

$$V = \int_0^1 \left[\left(\frac{\partial u}{\partial t} \right)^2 + \left(\frac{\partial u}{\partial x} \right)^2 \right] dx \tag{9.4}$$

It is immediately clear that, for $\gamma = 1$,

$$V \leq \gamma \rho^2 \tag{9.5}$$

so that the functional admits an upper bound, which, since ρ is really a function of time, can be made as small as desirable (at $t = 0$) by restricting the size of the initial disturbance. Further, we can rewrite equation (9.4) as

$$V = \int_0^1 \left[\left(\frac{\partial u}{\partial t} \right)^2 + (1 - c) \left(\frac{\partial u}{\partial x} \right)^2 + c \left(\frac{\partial u}{\partial x} \right)^2 \right] dx \tag{9.6}$$

which in view of the inequality (4.16) can be reckoned as

$$V \geq \int_0^1 \left[\left(\frac{\partial u}{\partial t} \right)^2 + (1 - c) \left(\frac{\partial u}{\partial x} \right)^2 + c \pi^2 u^2 \right] dx \tag{9.7}$$

If we now choose, for example, $c = 1/\pi^2$, then clearly

$$V \geq \alpha \rho^2 \tag{9.8}$$

if we also choose $\alpha = \dfrac{\pi^2 - 1}{\pi^2}$.

Finally, we note that since we are dealing with a conservative system, and since V is directly proportional to the total energy, we expect that $\dot{V} = 0$. Thus

$$
\begin{aligned}
\frac{\mathrm{d}V}{\mathrm{d}t} &= 2 \int_0^1 \left[\frac{\partial u}{\partial t} \frac{\partial^2 u}{\partial t^2} + \frac{\partial u}{\partial x} \frac{\partial^2 u}{\partial x \partial t} \right] \mathrm{d}x \\
&= 2 \int_0^1 \left[\frac{\partial u}{\partial t} \frac{\partial^2 u}{\partial x^2} + \frac{\partial u}{\partial x} \frac{\partial^2 u}{\partial t \partial x} \right] \mathrm{d}x \\
&= 2 \left[\frac{\partial u}{\partial x} \frac{\partial u}{\partial t} \right]_0^1 = 0
\end{aligned}
\tag{9.9}
$$

where we have used the differential equation (9.1), an integration by parts, and the boundary conditions (9.2) to achieve the final result.

Thus, by equations (9.5), (9.8) and (9.9), we see that we have met the requirements for the stability of the equilibrium position of the one-dimensional wave equation. The result can also be extended to the three-dimensional wave equation, as multi-dimensional inequalities analogous to equations (4.16) to (4.18) are available [4].

9.2 A hydrodynamics problem

Pritchard has given the following example, which is based on a mathematical model of turbulent flow in a channel [14]. The differential equation is

$$\frac{\partial u}{\partial t} = \frac{1}{R} \frac{\partial^2 u}{\partial x^2} + u - 2u \frac{\partial u}{\partial x} - Ru \int_0^1 u^2 \, \mathrm{d}x \tag{9.10}$$

subject to the boundary conditions

$$u(0) = u(1) = 0 \tag{9.11}$$

Here $u(x, t)$ represents the velocity of a turbulent disturbance, and the turbulence vanishes for $u = 0$. The parameter R is the Reynolds number.

We choose here the same functional for both the Lyapunov functional and the square of the metric. Thus let

$$V = \rho^2 = \int_0^1 u^2 \, \mathrm{d}x \tag{9.12}$$

Thus we have in a very simple fashion obtained a positive definite functional that admits an upper bound, with respect to the metric ρ. Now in evaluating the time rate of change we will examine first the linear problem, given by

$$\frac{\partial u}{\partial t} = \frac{1}{R} \frac{\partial^2 u}{\partial x^2} + u \tag{9.13}$$

Then by differentiating the functional (9.12) and substituting from equation (9.13) we find

$$\begin{aligned}
\frac{\mathrm{d}V}{\mathrm{d}t} &= 2 \int_0^1 u \left(\frac{1}{R} \frac{\partial^2 u}{\partial x^2} + u \right) \mathrm{d}x \\
&= 2 \int_0^1 \left[u^2 - \frac{1}{R} \left(\frac{\partial u}{\partial x} \right)^2 \right] \mathrm{d}x
\end{aligned} \tag{9.14}$$

where we have also noted the boundary conditions (9.11). By applying the inequality (4.16) we can obtain the result

$$\frac{\mathrm{d}V}{\mathrm{d}t} \leq 2 \int_0^1 \left(1 - \frac{\pi^2}{R} \right) u^2 \, \mathrm{d}x$$

or

$$\frac{\mathrm{d}V}{\mathrm{d}t} \leq -2 \left(\frac{\pi^2}{R} - 1 \right) V \tag{9.15}$$

Thus for $R < \pi^2$, we can state that the solution $u(x, t)$ is asymptotically stable.

For the nonlinear problem, the time derivative of the Lyapunov functional is

170

$$\frac{dV}{dt} = 2 \int_0^1 \left[\frac{1}{R} \frac{\partial^2 u}{\partial x^2} + u - 2u\frac{\partial u}{\partial x} - Ru \int_0^1 u^2 \, dx \right] u \, dx$$

which can also be written as

$$\frac{dV}{dt} = 2 \int_0^1 \left[\frac{u}{R} \frac{\partial^2 u}{\partial x^2} + u^2 - \tfrac{2}{3} \frac{\partial(u^3)}{\partial x} - Ru^2 \int_0^1 u^2 \, dx \right] dx \qquad (9.16)$$

Now the third term in the integrand vanishes because of the boundary conditions, while the fourth term is seen to be

$$- R \int_0^1 u^2 \, dx \int_0^1 u^2 \, dx = - RV^2 \qquad (9.17)$$

Then, treating the first term as before, we finally obtain

$$\frac{dV}{dt} \leq 2RV \left[\frac{1}{R} \left(1 - \frac{\pi^2}{R} \right) - V \right] \qquad (9.18)$$

Thus the nonlinear analysis yields the same result as the linear analysis, that is, the solution is asymptotically stable for $R < \pi^2$.

9.3 Linear anisotropic elasticity

We consider now the stability of the equilibrium position $u_i(x_j) = 0$ of a three-dimensional body made of an anisotropic, linear elastic material. The governing equation of motion is taken as

$$\frac{\partial^2 u_i}{\partial t^2} = C_{ijkl} \frac{\partial^2 u_l}{\partial x_j \partial x_k} \quad \text{in } B \qquad (9.19)$$

subject to the boundary condition

$$u_i = 0 \text{ on the surface of } B \qquad (9.20)$$

The term C_{ijkl} is a fourth-order tensor, having the symmetry properties

$$C_{ijkl} = C_{jikl} = C_{ijlk} = C_{klij} \qquad (9.21)$$

and also being such that for any second order tensor t_{ij} [4, 5]

$$\int_B C_{ijkl}\, t_{ij}t_{kl}\, \mathrm{d}B \geq C_0 \int_B t_{ij}\, t_{ij}\, \mathrm{d}B \tag{9.22}$$

We also assume that this tensor has been rendered dimensionless, as have the spatial and temporal coordinates.

For a Lyapunov functional we choose twice the total energy functional,

$$V = \int_B \left[\frac{\partial u_i}{\partial t}\, \frac{\partial u_i}{\partial t} + C_{ijkl}\, \frac{\partial u_i}{\partial x_j}\, \frac{\partial u_k}{\partial x_l} \right] \mathrm{d}B \tag{9.23}$$

for which it is easy to demonstrate that $\mathrm{d}V/\mathrm{d}t = 0$, either by straightforward manipulation or by the usual arguments tenable for a conservative system.

For the metric we choose the functional

$$\rho^2 = \int_B \left[\frac{\partial u_i}{\partial t}\, \frac{\partial u_i}{\partial t} + u_i u_i + \frac{\partial u_i}{\partial x_j}\, \frac{\partial u_i}{\partial x_j} \right] \mathrm{d}B \tag{9.24}$$

Upon comparing equations (9.23) and (9.24) it is clear that if a positive constant γ is chosen such that

$$\gamma = \max\, \{1,\, C_{ijkl}\} > 0 \tag{9.25}$$

then the functional (9.23) is bounded from above by the metric (9.24) as

$$V \leq \gamma \rho^2 \tag{9.26}$$

Now if we apply the inequality (9.22) to the Lyapunov functional, we find that

$$V \geq \int_B \left[\frac{\partial u_i}{\partial t}\, \frac{\partial u_i}{\partial t} + C_0\, \frac{\partial u_i}{\partial x_j}\, \frac{\partial u_i}{\partial x_j} \right] \mathrm{d}B$$

$$\geq \int_B \left[\frac{\partial u_i}{\partial t}\, \frac{\partial u_i}{\partial t} + (C_0 - C)\, \frac{\partial u_i}{\partial x_j}\, \frac{\partial u_i}{\partial x_j} + C\, \frac{\partial u_i}{\partial x_j}\, \frac{\partial u_i}{\partial x_j} \right] \mathrm{d}B \tag{9.27}$$

Now there are also two- and three-dimensional generalizations of the inequality (4.16), depending on the boundary condition (9.20) and upon an

172

appropriate Helmholtz-type eigenvalue problem, for which we can write that [4]

$$\int_B \frac{\partial u_i}{\partial x_j}\frac{\partial u_i}{\partial x_j}\,\mathrm{d}B \geq \lambda \int_B u_i u_i\,\mathrm{d}B \tag{9.28}$$

Thus the inequality (9.27) can be put into the form

$$V \geq \int_B \left[\frac{\partial u_i}{\partial t}\frac{\partial u_i}{\partial t} + (C_0 - C)\frac{\partial u_i}{\partial x_j}\frac{\partial u_i}{\partial x_j} + \lambda C u_i u_i\right]\mathrm{d}B \tag{9.29}$$

Then if we choose a positive number α such that [1]

$$\alpha = \min\{1,\, C_0 - C,\, \lambda C\} > 0 \tag{9.30}$$

it follows that our Lyapunov functional (9.23) is positive definite with respect to the metric (9.24), i.e.,

$$V \geq \alpha \rho^2 \tag{9.31}$$

The papers of Knops and Wilkes [4] and Knops and Payne [5] discuss this and related problems in a slightly different fashion, and they also explore the consequences of using different metrics, particularly those of the supremum type, e.g., equation (4.14). It is of interest to note that for those types of metrics, for hyperbolic systems such as equation (9.19), the above stability criteria cannot always be met. This is due to the phenomenon of 'focusing', and is demonstrated by, for example, locally unbounded displacements or strains. We have already discussed this possibility and its consequences for elastic stability definitions in Section 4.3.

9.4   An aeroelastic problem

As pointed out earlier in the chapter, the aeroelastic problem of determining the stability of a pre-stressed plate in an aerodynamic flow has been the

---

[1] Alternatively, as C is arbitrary, we could choose $C = 1/\lambda$, and take $\alpha = \min\{1, C_0 - \lambda^{-1}\}$.

subject of a number of Lyapunov functional investigations. We will present here the results of a problem considered by Plaut [6].

Now we state at the outset that although aeroelasticity is a very interesting subject, we are interested here only in the (applied) mathematics of the stability of a solution to a nonlinear differential equation. For more extensive coverage of the physics, the reader should go to Refs. [6-9], and the classic texts of Fung [24] and Bisplinghoff and Ashley [25].

In dimensionless form, the deflection $y(x, t)$ of a 'one-dimensional' plate is obtained as the solution of the nonlinear partial integro-differential equation,

$$\frac{\partial^2 y}{\partial t^2} + \frac{\partial^4 (y - y_0)}{\partial x^4} + 2 \frac{\partial^2 y}{\partial x^2} \int_0^1 \left[\left(\frac{\partial y_0}{\partial x} \right)^2 - \left(\frac{\partial y}{\partial x} \right)^2 \right] dx$$

$$- f \frac{\partial^2 (y - y_0)}{\partial x^2} + \Omega \frac{\partial y}{\partial x} + \Gamma \frac{\partial y}{\partial t} = 0 \tag{9.32}$$

Here y_0 represents the initial deviation from straightness of the cylindrically bent plate, x is the axial coordinate, f an in-plane force taken as positive in tension, and Ω and Γ are flow parameters due to the fluid motion. The fluid parameters are both dependent on the local Mach number of the flow, i.e.,

$$\Omega \sim \frac{M^2}{(M^2 - 1)^{1/2}}, \quad \Gamma \sim \frac{M(M^2 - 2)}{(M^2 - 1)^{3/2}} \tag{9.33}$$

and the Mach number itself is assumed to be in the supersonic range defined by $M^2 > 2$. Finally, we note that the boundary conditions at $x = 0,1$ will be taken to correspond to a pinned column or plate, for which the static buckling load would be $f = -\pi^2$.

Now the *equilibrium configuration* is defined by the static deflection $y_s(x)$, with steady flow, i.e.,

$$\frac{\partial^4 (y_s - y_0)}{\partial x^4} + 2 \frac{\partial^2 y_s}{\partial x^2} \int_0^1 \left[\left(\frac{\partial y_0}{\partial x} \right)^2 - \left(\frac{\partial y_s}{\partial x} \right)^2 \right] dx$$

$$- f \frac{\partial^2 (y_s - y_0)}{\partial x^2} + \Omega \frac{\partial y_s}{\partial x} = 0 \tag{9.34}$$

We now introduce the local dynamic disturbance $w(x, t)$, defined by

$$w(x, t) = y(x, t) - y_s(x) \tag{9.35}$$

Thus we will investigate the stability of the equilibrium configuration defined by $w(x, t) = 0$, and we can accomplish this by considering the linearized version of equation (9.32). Thus we will consider the partial integro-differential equation

$$\frac{\partial^2 w}{\partial t^2} + \frac{\partial^4 w}{\partial x^4} - f\frac{\partial^2 w}{\partial x^2} + 2\frac{\partial^2 w}{\partial x^2}\int_0^1\left[\left(\frac{\partial y_0}{\partial x}\right)^2 - \left(\frac{\partial y_s}{\partial x}\right)^2\right]dx$$

$$- 4\frac{\partial^2 y_s}{\partial x^2}\int_0^1\frac{\partial w}{\partial x}\frac{\partial y_s}{\partial x}\,dx + \Omega\frac{\partial w}{\partial x} + \Gamma\frac{\partial w}{\partial t} = 0 \tag{9.36}$$

subject to the boundary conditions

$$w = \frac{\partial^2 w}{\partial x^2} = 0 \quad \text{for} \quad x = 0, 1 \quad \text{for all} \quad t > 0. \tag{9.37}$$

As the metric we choose the same one used in the column analysis in Chapter 4, so that

$$\rho^2 = \int_0^1\left[\left(\frac{\partial w}{\partial t}\right)^2 + w^2 + \left(\frac{\partial w}{\partial x}\right)^2 + \left(\frac{\partial^2 w}{\partial x^2}\right)^2\right]dx \tag{9.38}$$

In preparing to choose a Lyapunov functional for this problem, we note that the energy functional may not be appropriate as the airstream may be either adding energy to the vibrating panel, or carrying energy away, or both, at different places on the panel and at different times during the motion. Nevertheless, as a starting point, we will take twice the energy of the linearized system as our functional [6]:

$$V(w) = \int_0^1\left[\left(\frac{\partial w}{\partial t}\right)^2 + \left(\frac{\partial^2 w}{\partial x^2}\right)^2 + f\left(\frac{\partial w}{\partial x}\right)^2\right]dx$$

$$+ 4\left\{\int_0^1\frac{\partial w}{\partial x}\frac{\partial y_s}{\partial x}\,dx\right\}^2 - 2\int_0^1\left(\frac{\partial w}{\partial x}\right)^2 dx\int_0^1\left[\left(\frac{\partial y_0}{\partial x}\right)^2 - \left(\frac{\partial y_s}{\partial x}\right)^2\right]dx \tag{9.39}$$

It is a straightforward task to differentiate the functional (9.39), and to use the differential equation (9.36) and the boundary conditions (9.37) to find the time rate of change of the functional, i.e.,

$$\frac{\mathrm{d}V(w)}{\mathrm{d}t} = -2\int_0^1\left[\Gamma\left(\frac{\partial w}{\partial t}\right)^2 + \Omega\,\frac{\partial w}{\partial t}\,\frac{\partial w}{\partial x}\right]\mathrm{d}x \tag{9.40}$$

As expected, this energy-rate term is not of definite sign.

However, we observe that if we could add a term

$$-2C^2\int_0^1\left(\frac{\partial w}{\partial x}\right)^2\mathrm{d}x$$

to the right side of equation (9.40), with an appropriately chosen constant C^2, we could find a negative definite time rate of change.

Now it turns out that if we considered the functional

$$V'(w) = \int_0^1\left[\frac{1}{2}\Gamma^2\,w^2 + \Gamma w\,\frac{\partial w}{\partial t}\right]\mathrm{d}x \tag{9.41}$$

we could obtain its time derivative as

$$\frac{\mathrm{d}V'(w)}{\mathrm{d}t} = 2\Gamma\int_0^1\left(\frac{\partial w}{\partial t}\right)^2\mathrm{d}x - \Gamma V(w) \tag{9.42}$$

and clearly $V(w)$ contains a term of the type sought. Thus we shall choose as our Lyapunov functional

$$V''(w) = V(w) + V'(w)$$

$$= \int_0^1\left[\left(\frac{\partial w}{\partial t}\right)^2 + \left(\frac{\partial^2 w}{\partial x^2}\right)^2 + (\varLambda - \pi^2)\left(\frac{\partial w}{\partial x}\right)^2 + \frac{1}{2}\Gamma^2 w^2\right.$$

$$\left. + \Gamma w\,\frac{\partial w}{\partial t}\right]\mathrm{d}x + 4\left\{\int_0^1\frac{\partial w}{\partial x}\,\frac{\partial y_s}{\partial x}\,\mathrm{d}x\right\}^2 \tag{9.43}$$

where we have introduced the parameter $\varLambda$, defined as

$$\varLambda = \pi^2 + f - 2\int_0^1\left[\left(\frac{\partial y_0}{\partial x}\right)^2 - \left(\frac{\partial y_s}{\partial x}\right)^2\right]\mathrm{d}x \tag{9.44}$$

The time derivative of our new functional is then obtained as

$$\frac{\mathrm{d}V''(w)}{\mathrm{d}t} = -\int_0^1 \left[\Gamma \left(\frac{\partial w}{\partial t}\right)^2 + \Gamma \left(\frac{\partial^2 w}{\partial x^2}\right)^2 + (\Lambda - \pi^2)\, \Gamma \left(\frac{\partial w}{\partial x}\right)^2 \right.$$

$$\left. + 2\Omega\, \frac{\partial w}{\partial x}\, \frac{\partial w}{\partial t} \right] \mathrm{d}x - 4\Gamma \left\{ \int_0^1 \frac{\partial w}{\partial x}\, \frac{\partial y_s}{\partial x}\, \mathrm{d}x \right\}^2 \tag{9.45}$$

In order to apply the conditions of the stability theorem, which we are now ready to do, we must first show that $V''(w)$ is positive definite with respect to the metric (9.38). From equation (9.43) we can see that

$$V''(w) \geq \int_0^1 \left[\left(\frac{\partial w}{\partial t}\right)^2 + \left(\frac{\partial^2 w}{\partial x^2}\right)^2 + (\Lambda - \pi^2)\left(\frac{\partial w}{\partial x}\right)^2 \right.$$

$$\left. + \frac{1}{2}\,\Gamma^2 w^2 + \Gamma w\, \frac{\partial w}{\partial t} \right] \mathrm{d}x \tag{9.46}$$

Also, since

$$\left(\frac{\partial w}{\partial t}\right)^2 + \Gamma w\, \frac{\partial w}{\partial t} + \frac{1}{2}\Gamma^2 w^2 = \left(\frac{\partial w}{\partial t} + \frac{1}{2}\Gamma w\right)^2 + \frac{1}{4}\Gamma^2 w^2 > 0$$

it is not difficult to find two numbers α_1, α_2, both less than unity, such that

$$\left(\frac{\partial w}{\partial t}\right)^2 + \Gamma w\, \frac{\partial w}{\partial t} + \frac{1}{2}\,\Gamma^2 w^2 \geq \alpha_1 \left(\frac{\partial w}{\partial t}\right)^2 + \alpha_2\, \Gamma^2 w^2 \tag{9.47}$$

The only requirement is that

$$2(1 - \alpha_1)(1 - 2\alpha_2) = 1 \tag{9.48}$$

Thus, for example, if $\alpha_1 = \frac{1}{4}$, then $\alpha_2 = \frac{1}{6}$. Then we have

$$V''(w) \geq \int_0^1 \left[\alpha_1 \left(\frac{\partial w}{\partial t}\right)^2 + \left(\frac{\partial^2 w}{\partial x^2}\right)^2 + (\Lambda - \pi^2)\left(\frac{\partial w}{\partial x}\right)^2 \right.$$

$$\left. + \alpha_2\, \Gamma^2 w^2 \right] \mathrm{d}x \tag{9.49}$$

A comparison of equations (9.49) and (9.38) then makes it quite clear that it is easy to pick a value of α, assuming of course that $\Lambda - \pi^2 > 0$, to insure the positivity of $V''(w)$. For example, choose

$$\alpha = \min\{\alpha_1, \Lambda - \pi^2, \alpha_2 \Gamma^2, 1\} > 0 \tag{9.50}$$

To prove that $V''(w)$ is bounded, we note first that by the Schwartz inequality

$$\left\{ \int_0^1 \frac{\partial w}{\partial x} \frac{\partial y_s}{\partial x} \, dx \right\}^2 \leq C_s^2 \int_0^1 \left(\frac{\partial w}{\partial x} \right)^2 dx \tag{9.51}$$

where

$$C_s^2 = \int_0^1 \left(\frac{\partial y_s}{\partial x} \right)^2 dx \tag{9.52}$$

Further, note that $\Gamma w \dfrac{\partial w}{\partial t} \leq \dfrac{1}{2}\left[\left(\dfrac{\partial w}{\partial t} \right)^2 + \Gamma^2 w^2 \right]$. Then, from these results and equation (9.43), it follows that

$$V''(w) \leq \int_0^1 \left[\frac{3}{2} \left(\frac{\partial w}{\partial t} \right)^2 + \left(\frac{\partial^2 w}{\partial x^2} \right)^2 + (4C_s^2 + \Lambda - \pi^2) \left(\frac{\partial w}{\partial x} \right)^2 \right.$$

$$\left. + \ \Gamma^2 w^2 \right] dx \tag{9.53}$$

Hence if we choose

$$\gamma = \max\left\{ \frac{3}{2}, \ 4C_s^2 + \Lambda - \pi^2, \ \Gamma^2 \right\} > 0 \tag{9.54}$$

we may be assured that $V''(w)$ admits an upper bound with respect to the metric (9.43).

Now if we examine the time derivative of our functional, equation (9.45), we see immediately that

$$\frac{dV''(w)}{dt} \leq - \int_0^1 \left[\Gamma \left(\frac{\partial w}{\partial t} \right)^2 + \Gamma \left(\frac{\partial^2 w}{\partial x^2} \right)^2 \right.$$

$$+ \Gamma(\Lambda - \pi^2) \left(\frac{\partial w}{\partial x}\right)^2 + 2\Omega \frac{\partial w}{\partial x} \frac{\partial w}{\partial t}\right] \mathrm{d}x \tag{9.55}$$

In view of the inequality (4.18), it also follows that

$$\frac{\mathrm{d}V''(w)}{\mathrm{d}t} \leq - \int_0^1 \left[\Gamma \left(\frac{\partial w}{\partial t}\right)^2 + 2\Omega \frac{\partial w}{\partial x} \frac{\partial w}{\partial t} + \Gamma\Lambda \left(\frac{\partial w}{\partial x}\right)^2 \right] \mathrm{d}x \tag{9.56}$$

But this integrand is a quadratic form which will be definite if $\Gamma\Lambda$ and $\Gamma^2\Lambda - \Omega^2$ have the same sign. For our purposes, then, since Γ is positive by definition, we require that

$$\Lambda > 0 \quad \text{and} \quad \Omega^2 < \Gamma^2\Lambda \tag{9.57}$$

for which $\mathrm{d}V''(w)/\mathrm{d}t < 0$, and stability is assured.

Thus we have obtained stability criteria for a pre-stressed, imperfect, cylindrical plate, subject to a fluid loading on one side. Note that we have not had to specify the imperfection, $y_0(x)$, nor did we have to solve for the static equilibrium deflection, $y_s(x)$, to obtain the stability boundaries.

We might note in closing this section that we have conducted what is often called a flutter analysis to determine a flutter boundary. Flutter is defined as a mode of instability whereby a statically stable (or steady) state becomes unstable through oscillatory motion of a self-perpetuating type.

9.5 Instability of an Euler column

As a final example of an investigation whose primary focus is on the stability of a structural member, we will return to the Euler column considered in Chapter 4. Here, however, we will concentrate on the instability theorem of Movchan (Theorem 4.2), and show that the pinned column is unstable [1] for $p > \pi^2$. Plaut has performed this calculation for the clamped-clamped column [6].

---

[1] As is hopefully clear from the results of Chapter 6, we don't really mean this as a general proposition. This will be true only for the linearized theory, as outlined in Sections 4.1, 6.1 and 6.2.

Following the prescription of Theorem 4.2, we need first to find an element u° for which $0 < \rho(u^\circ, 0) < \delta$ and for which there exists a positive definite functional, $W(u^\circ) > 0$.

Let

$$W(u) = \begin{cases} -V_1(u)g(u) & \text{if } V_1(u) < 0 \\ 0 & \text{if } V_1(u) > 0 \end{cases} \tag{9.58}$$

where $V_1(u)$ is given by equation (4.6), and

$$g(u) = \int_0^1 wv\,\mathrm{d}x \tag{9.59}$$

As in Chapter 4, $v = \partial w/\partial t$.

Now let

$$u^\circ = \begin{pmatrix} c_1 \sin \pi x \\ c_2 \sin \pi x \end{pmatrix} \tag{9.60}$$

whence it follows that

$$\rho^2(u^\circ, 0) = \frac{1}{2}[c_2^2 + (1 + \pi^2 + \pi^4)c_1^2] \tag{9.61}$$

$$V_1(u^\circ) = \frac{1}{2}[c_2^2 + (\pi^4 - \pi^2 p)c_1^2] \tag{9.62}$$

$$g(u^\circ) = \frac{1}{2}c_1 c_2 \tag{9.63}$$

Then, for $p > \pi^2$, we need only choose c_1 and c_2 so that $\rho(u^\circ, 0) < \delta$, where δ is given, and so that $V_1(u^\circ) < 0$ and $g(u^\circ) > 0$. For example, choose

$$c_1^2 = \frac{\delta^2}{M^2}, \quad c_2^2 = (p - \pi^2)\frac{\delta^2}{M^2} \tag{9.64}$$

where the number M can be taken to be as large as is desired.

Now we must demonstrate that $|W(u)|$ is bounded from above by $\rho_1^4(u, 0)$. We have already demonstrated that

$$|V_1(u)| \le \gamma \rho_1^2(u, 0) \quad \text{for } \gamma = \max\{1, |p|\} > 0 \tag{9.65}$$

Further, since $(v - w)^2 \geq 0$,

$$g(\boldsymbol{u}) \leq \frac{1}{2} \int_0^1 [v^2 + w^2]\,\mathrm{d}x \leq \rho_1^2(\boldsymbol{u}, \boldsymbol{0}) \tag{9.66}$$

Hence we can combine equations (9.65) and (9.66) with the definition of $W(\boldsymbol{u})$ to find that

$$|\,W(\boldsymbol{u})\,| = |\,V_1(\boldsymbol{u})\,|\,|\,g(\boldsymbol{u})\,| \leq \gamma\rho_1^4(\boldsymbol{u}, \boldsymbol{0}), \gamma > 0 \tag{9.67}$$

As the final step in the instability proof, we must demonstrate that whenever $W(\boldsymbol{u}) > 0$, we also have $\mathrm{d}W(\boldsymbol{u})/\mathrm{d}t > 0$.

For some positive time $t = t^*$, assume that $W(\boldsymbol{u})$ is positive. By the definition (9.58) it follows that $V_1(\boldsymbol{u})$ is negative there, and for all time $t \geq 0$. This is true because $V_1(\boldsymbol{u})$ is an energy functional, hence $\mathrm{d}V_1(\boldsymbol{u})/\mathrm{d}t = 0$. And it also follows that

$$\frac{\mathrm{d}W_1(\boldsymbol{u})}{\mathrm{d}t} = -V_1(\boldsymbol{u})\,\frac{\mathrm{d}g(\boldsymbol{u})}{\mathrm{d}t} \tag{9.68}$$

Now

$$\frac{\mathrm{d}g(\boldsymbol{u})}{\mathrm{d}t} = \int_0^1 \left[v^2 + w\,\frac{\partial v}{\partial t}\right]\mathrm{d}x$$

$$= \int_0^1 \left[v^2 - pw\,\frac{\partial^2 w}{\partial x^2} - w\,\frac{\partial^4 w}{\partial x^4}\right]\mathrm{d}x$$

$$= 2\int_0^1 v^2\,\mathrm{d}x - V_1(\boldsymbol{u}) \tag{9.69}$$

where we have used the differential equation (4.4) and the pinned-pinned boundary conditions.

Hence, substituting from equation (9.69),

$$\frac{\mathrm{d}W_1(\boldsymbol{u})}{\mathrm{d}t} = V_1^2(\boldsymbol{u}) - 2V_1(\boldsymbol{u})\int_0^1 v^2\,\mathrm{d}x > 0 \tag{9.70}$$

since, of course, $V_1(\boldsymbol{u}) < 0$ by assumption.

Thus we have shown that, according to the linear theory, $p < \pi^2$ is both necessary and sufficient for stability of the trivial (equilibrium) state with respect to the metric ρ_1.

9.6   A bounding example

In this section we shall outline an approach to bounding the displacements of a beam-column, using Lyapunov functionals. First we note a result given by Freund and Plaut [15] regarding the displacements of a beam, based on optimizing the Green's function for a particular beam. They determined that

$$w^2(x, t) \leq \frac{1}{k^2} \int_0^1 \left(\frac{\partial^2 w}{\partial x^2}\right)^2 \mathrm{d}x \tag{9.71}$$

where $k^2 = 6.928$ for a pinned-pinned beam, $k^2 = 13.86$ for a clamped-clamped beam, and $k^2 = 10.09$ for a clamped-pinned beam.

Now the equation for the dynamic response of a beam-column, initally imperfect, can by given in dimensionless form as (9.16)

$$\frac{\partial^4 w}{\partial x^4} + p(t)\frac{\partial^2 w}{\partial x^2} + \frac{\partial^2 w}{\partial t^2} = r(x, t) \tag{9.72}$$

where

$$r(x, t) = q(x, t) - p(t)\frac{\partial^2 w_0}{\partial x^2} \tag{9.73}$$

Here $q(x, t)$ is an applied transverse load.

Consider now the Lyapunov functional consisting of twice the sum of the kinetic energy and the strain energy due to bending, that is, let

$$V = \int_0^1 \left[\left(\frac{\partial w}{\partial t}\right)^2 + \left(\frac{\partial^2 w}{\partial x^2}\right)^2\right] \mathrm{d}x \tag{9.74}$$

Then

$$\frac{\mathrm{d}V}{\mathrm{d}t} = 2\int_0^1 \left[\frac{\partial w}{\partial t}\frac{\partial^2 w}{\partial t^2} + \frac{\partial^2 w}{\partial x^2}\frac{\partial^3 w}{\partial t \partial x^2}\right]\mathrm{d}x$$

$$= 2\left[\frac{\partial^2 w}{\partial x^2}\frac{\partial^2 w}{\partial t \partial x} - \frac{\partial^3 w}{\partial x^3}\frac{\partial w}{\partial t}\right]_0^1$$

$$+ 2\int_0^1 \frac{\partial w}{\partial t}\left[\frac{\partial^2 w}{\partial t^2} + \frac{\partial^4 w}{\partial x^4}\right]dx \tag{9.75}$$

where the second term in the integrand of dV/dt has been integrated by parts, twice. For any of the three sets of boundary conditions mentioned earlier, the bracketed term in equation (9.75) vanishes. Now making use of the differential equation (9.72) we see that

$$\frac{dV}{dt} = 2\int_0^1\left[r(x, t) - p(t)\frac{\partial^2 w}{\partial x^2}\right]\frac{\partial w}{\partial t}dx$$

$$= 2\int_0^1 r(x, t)\frac{\partial w}{\partial t}\,dx - 2p(t)\int_0^1 \frac{\partial^2 w}{\partial x^2}\frac{\partial w}{\partial t}dx \tag{9.76}$$

At this point we note that

$$\pm 2\int_0^1 \frac{\partial^2 w}{\partial x^2}\frac{\partial w}{\partial t}\,dx \le \int_0^1\left[\left(\frac{\partial^2 w}{\partial x^2}\right)^2 + \left(\frac{\partial w}{\partial t}\right)^2\right]dx = V \tag{9.77}$$

and we also note by Schwartz's inequality that

$$\int_0^1 r(x, t)\frac{\partial w}{\partial t}\,dx \le \left\{\int_0^1 r^2(x, t)\,dx\right\}^{1/2}\left\{\int_0^1 \left(\frac{\partial w}{\partial t}\right)^2 dx\right\}^{1/2}$$

$$\le v(t)\,V^{1/2} \tag{9.78}$$

where

$$v(t) \equiv \left\{\int_0^1 r^2(x, t)\,dx\right\}^{1/2} \tag{9.79}$$

Thus, by combining the results (9.76), (9.77) and (9.78), we find that

$$\frac{dV}{dt} \le |\,p(t)\,|\,V + 2v(t)\,V^{1/2} \tag{9.80}$$

If we make the transformation $V = U^2$, we get

$$2U \, \frac{dU}{dt} \leq |p(t)| \, U^2 + 2v(t) \, U(t)$$

or

$$\frac{dU}{dt} \leq \frac{1}{2} |p(t)| \, U(t) + v(t) \tag{9.82}$$

This inequality can be integrated, and it yields, for a beam-column initially at rest,

$$U(t) \leq \int_0^t v(t') \exp \left\{ \tfrac{1}{2} \int_{t'}^t |p(t'')| \, dt'' \right\} dt' \tag{9.83}$$

Then, if we denote our maximum displacement in the beam as $w_{\max}$, we can combine equations (9.71), (9.74) and (9.83) to yield the bound

$$w_{\max} \leq \frac{1}{k} U(t) \leq \frac{1}{k} \int_0^t v(t') \exp \left\{ \tfrac{1}{2} \int_{t'}^t |p(t'')| \, dt'' \right\} dt' \tag{9.84}$$

Thus the maximum value of the displacement can be bounded in terms of a constant that depends on the boundary conditions, and upon the axial and lateral loads and the initial curvature. Further applications are given in Refs. [16-19].

9.7   Summary

We have given in this chapter a précis of current research on the applications of Lyapunov functionals to stability investigations in continuous systems. We have included problems of a classical mathematical nature, as well as current work in structural mechanics. Nevertheless, this is only an introduction, and the reader is encouraged to read the works cited below for his own edification.

References

[1] V. I. Zubov: *Methods of A. M. Lyapunov and their Applications*. Leningrad, 1957; English translation, P. Noordhoff Ltd., Groningen, Holland, 1964.

[2] A. A. Movchan: The Direct Method of Liapunov in Stability Problems of Elastic Systems. *Prikladnaya Matematika i Mekhanika*, Vol. 23, 1959, pp. 483-493; also, *Applied Mathematics and Mechanics*, Vol. 23, No. 3, 1959, pp. 686-700.

[3] A. A. Movchan: Stability of Processes with Respect to Two Metrics. *Prikladnaya Matematika i Mekhanika*, Vol. 24, 1960, pp. 988-1001; also *Applied Mathematics and Mechanics*, Vol. 24, No. 6, 1960, pp. 1506-1524.

[4] R. J. Knops and E. W. Wilkes: On Movchan's Theorems for Stability of Continuous Systems. *International Journal of Engineering Science*, Vol. 4, 1966, pp. 303-329.

[5] R. J. Knops and L. E. Payne: Stability in Linear Elasticity. *International Journal of Solids and Structures*, Vol. 4, 1968, pp. 1233-1242.

[6] R. H. Plaut: A Study of the Dynamic Stability of Continuous Elastic Systems by Liapunov's Direct Method. *University of California, Berkeley* Report No. AM-67-3, May 1967.

[7] P. C. Parks: A Stability Criterion for Panel Flutter Problem via the Second Method of Liapunov. *AIAA Journal*, Vol. 3, No. 9, 1965, pp. 1764-1766.

[8] P. K. C. Wang: Stability Analysis of a Simplified Flexible Vehicle via Lyapunov's Direct Method. *AIAA Journal*, Vol. 4, No. 1, 1966, pp. 175-178.

[9] P. K. C. Wang: Stability Analysis of Elastic and Aeroelastic Systems via Lyapunov's Direct Method. *Journal of the Franklin Institute*, Vol. 281, No. 1, 1966, pp. 51-72.

[10] G. A. Hegemier: Stability of Cylindrical Shells Under Moving Loads by the Direct Method of Liapunov. *Journal of Applied Mechanics*, Vol. 34, No. 4, December 1967, p. 991.

[11] P. C. Parks and A. J. Pritchard: Stability Analysis in Structural Dynamics Using Liapunov Functionals. *Journal of Sound and Vibration*, Vol. 25, No. 4. December 1972, p. 609.

[12] E. F. Infante and R. H. Plaut: Stability of a Column Subjected to a Time-Dependent Axial Load. *AIAA Journal*, Vol. 7, No. 4, April 1969, p. 766.

[13] A. J. Berger and L. Lapidus: An Introduction to the Stability of Distributed Systems via a Liapunov Functional. *AIChE Journal*, Vol. 14, No. 4, July 1968, p. 558.

References

[14] A. J. PRITCHARD: On Nonlinear Stability Theory. *Quarterly of Applied Mathematics*, Vol. 27, No. 4, 1970, pp. 531-536.

[15] L. B. FREUND and R. H. PLAUT: An Energy-Displacement Inequality Applicable to Problems in the Dynamic Stability of Structures. *Journal of Applied Mechanics*, Vol. 38, No. 2, June 1971, p. 536.

[16] R. H. PLAUT: Displacement Bounds for Beam-Columns With Initial Curvature Subjected to Transient Loads. *International Journal of Solids and Structures*, Vol. 7, No. 9, September 1971, p. 1229.

[17] R. H. PLAUT and E. F. INFANTE: Bounds on Motion of Some Lumped and Continuous Dynamic Systems. *Journal of Applied Mechanics*, Vol. 39, No. 1, March 1972, p. 251.

[18] S. M. HOLZER: Response Bounds for Columns With Transient Loads. *Journal of Applied Mechanics*, Vol. 38, No. 1, March 1971, p. 157.

[19] S. M. HOLZER: Stability and Boundedness Via Liapunov's Direct Method. *Journal of the Engineering Mechanics Division*, ASCE, Vol. 98, No. EM5, October 1972, p. 1273.

[20] J. A. WALKER: Energy-like Liapunov Functionals for Linear Elastic Systems on a Hilbert Space. *Quarterly of Applied Mathematics*, Vol. 30, No. 4, January 1973, p. 465.

[21] J. A. WALKER: Liapunov Analysis of the Generalized Pflüger Problem. *Journal of Applied Mechanics*, Vol. 39, No. 4, December 1972, p. 935.

[22] J. A. WALKER: Stability of a Pin-Ended Bar in Torsion and Compression. *Journal of Applied Mechanics*, Vol. 40, No. 2, June 1973, p. 405.

[23] J. A. WALKER and M. W. DIXON: Stability of the General Plane Membrane Adjacent to a Supersonic Airstream. *Journal of Applied Mechanics*, Vol. 40, No. 2, June 1973, p. 395.

[24] Y. C. FUNG: *An Introduction to the Theory of Aeroelasticity*. Dover Publications, New York, 1969.

[25] R. L. BISPLINGHOFF and H. ASHLEY: *Principles of Aeroelasticity*. John Wiley, New York, 1962.

Index to authors

Subject index